# AEREALITY

## ALSO BY WILLIAM L. FOX

ESSAYS ON THE WORLD FROM ABOVE

# AEREALITY

## WILLIAM L. FOX

COUNTERPOINT
BERKELEY

Much of the section in Chapter 7 about the photographer David Maisel first appeared in an earlier version as an essay in *Oblivion*, published by Nazraeli Press in 2006.

For a full list of credits see page 337.

Members of Aboriginal communities are respectfully advised that some of the people mentioned in writing have passed away.

Library of Congress Cataloging-in-Publication Data
Fox, William L., 1949–
    Aereality : on the world from above / William L. Fox.
        p.cm.
    Includes bibliographical references.
    ISBN 978-1-58243-429-2
1. Aerial photography. I. Title.

    TR810.F688 2009
    778.3'5—dc22

                                    2008034647

Front cover design by David Bullen
Interior design by Elyse Strongin, Neuwirth & Associates, Inc.

Printed in the United States of America

COUNTERPOINT
2117 Fourth Street
Suite D
Berkeley, CA 94710
www.counterpointpress.com

Distributed by Publishers Group West

10   9   8   7   6   5   4   3   2   1

for Denis Cosgrove (1948-2008)

# CONTENTS

## PART 3:  Down Under

# LIST OF PLATES

# AEREALITY

# TWO PICTURES
# AND A STORY

**"Aréalité " is an almost completely obscured French word that curator** Thierry Davila from the Bordeaux Museum of Contemporary Art brought to my attention when I told him that I was writing about "the aerial imagination." He translates it as "air + reality," but to my ear it immediately fused "aerial" and "reality" in a near palindromic fashion, never mind that I couldn't find it in any number of French dictionaries. As I thought about "aereality," it brought to mind another term that I often reflect upon: "fundamental." This is a word that compounds two Latin terms, *fund* for that which is at the bottom or ground level, and *mens* for that which is in the mind. The overarching theme of my books, whether I am writing about art or science, in poetry or in prose, is how human cognition interacts with land in order to transform it into landscape. You can see why a word such as "fundamental" might be attractive to me, although I have never used it in a title.

When looking at the ground as much as I do, it is also natural to look up into the air, and then to wonder how the world would look from some higher vantage point. This is a book about how and why we perform such actions in imagination and in reality, and I thought it might be useful to resurrect an existing word for it, one that would keep me in two minds at once with my feet on the ground and my head in the clouds, as it were, while thinking about the ubiquity of aerial views in contemporary life. *Aereality* it is.

But before I had a title, I had an argument among two pictures and a story. The first image was a Neolithic wall-painting from Anatolia, the second an aerial photograph made over the Mojave Desert almost eight thousand years later; the story was one told to me by a geologist from Melbourne about Aborigines in Western Australia. It's not a coincidence that all three come from arid or semiarid places.

I spend much of my time tracking down artifacts made in large spaces such as deserts and the polar regions, extreme environments in which the brain struggles to create place. These are sites where human neurophysiology has more trouble functioning than in those that more closely resemble the temperate forests and savannas in which we mostly evolved. As a result, the traces we leave behind of our place-making struggle in arid lands are often more dramatic than those found in milder terrain. The artifacts, evidence of this transformational process, are an indication of how we overcome our natural cognitive limitations by deploying cultural adaptations such as cartography and art, two types of symbol systems that at times have been one and the same. Most of the artifacts I examine are visual byproducts of this difficult process of converting land into landscape, or terrain into territory. That's the consequence of vision being our dominant sense, 80 percent of everything we learn each day coming through our eyes. The artifacts also tend to be polyvalent, in part because they are the results of one kind of representational system being adapted to a new circumstance. Maps are paintings, paintings are strategic documents, documents read like poems, habitat becomes sculpture, sculpture erodes into an archeological ruin.

It doesn't hurt, either, that the layers of culture tend to be fewer and better preserved in deserts, be they hot or polar, than those in other places, and this allows their artifacts to be used more easily as vantage points from which to survey history. Some of the cultural traces are prehistoric, such as rock art, and some historic, such as maps and the exploration art of the seventeenth and eighteenth centuries. But most of the images I examine are contemporary ones made by the artists and scientists with whom I travel. It's exciting, however, that I would find two images widely separated in time to start a conversation.

While writing a book several years ago about the history of the artistic, cartographic, and scientific images of the most extreme environment on the planet, the Antarctic, I came across what some people call the oldest map and others the oldest landscape painting. It is also, as far as I can tell, the oldest known aerial representation of a place. Çatalhöyük was a Neolithic settlement in Anatolia (present-day Turkey), an agricultural and trading center that was continuously inhabited for 1400 years between 7400 and 6000 BC. Its flat-roofed mud-brick houses essentially shared walls so that there were neither streets nor front doors. To get in and out of the houses, people climbed an indoor ladder to a hole in the roof of the house (these holes serving also as smoke vents), and then walked across their neighbors' roofs to climb down another roof-hole. Situated near marshes that were slowly drying out in the increasingly arid climate, the people grew wheat and barley and peas, and traded obsidian with other towns. The site, which covers more than thirty-two acres, is still only 5 percent excavated, a process initiated in 1961–1965 by James Mellaart, a British archeologist, then resumed in 1993 and directed continuously since by Ian Hodder.

Every three or four generations, about eighty years on average, the wooden wall posts supporting the roof beams were pulled out, the upper walls were knocked down into the rooms, and the house was rebuilt on top of its own rubble, eventually creating a mound that rose sixty-five feet above the surrounding farmland. Between three and eight thousand people occupied each successive layer at Çatalhöyük, so in total between 50,000 at the least and 150,000 at the most. In each house at least one wall was covered in white plaster and painted with murals. Walls were plastered over annually, perhaps even seasonally, and redecorated. One wall that's been uncovered looks to have been repainted some 120 times. Hodder admits that there are numerous Neolithic sites that have been dated to be a thousand years older than Çatalhöyük, but adds that none of them have such sheer density of imagery.

One of the important facets of Çatalhöyük is that it was the first excavation to materially demonstrate the development of art from the Upper Paleolithic—the end of the Old Stone Age of hunter/gatherers—into the Neolithic age of early agriculture, then into the Chalcolithic and Bronze ages. At its lowest levels the decorations are relatively simple geometric patterns. As we move toward the top of the excavation and later periods, the patterns become increasingly complex and appear to imitate early woven textiles, such as kilim, a proposition supported by the fact that patterned fibers predate the paintings. Higher up in the deposits, naturalistic depictions of animals and people are added. In the middle of what I am describing here—as a linear narrative, which in reality is nowhere near as simple as a progressively more sophisticated set of styles—Mellaart found an extraordinary wall painting done along the north and east wall of a ritually decorated house.

This "volcano painting," a panoramic view done around 6200 BC, shows the town in planimetric (a plan view, as if seen from straight above) and the then active Hasan Dag volcano, its twin summits sixty miles away reaching 10,672 feet, in elevation (in profile as if seen from a horizontal view). The volcano went extinct around 2,000 BC, but during the time of Çatalhöyük it was a live source of that obsidian traded as far south as Jericho. How did someone make this image, which accurately represents the scale and relationship of houses to one another, two thousand years before the invention of writing and four thousand years before the invention of any known alphabet? How did they assume an aerial view, as if they were standing above a roughly gridded street map? There were no hills nearby Çatalhöyük, and although the residents apparently climbed up the volcano to obtain obsidian, the town was effectively invisible from that distance. Perhaps standing on the roofs, on the mound itself, helped inculcate an aerial view of the town—but it wouldn't have been a vertical view like this. And why make this composite image in plan and profile?

The image is not just unique in terms of its age, but also because it offers us a compound view of a landscape, the town laid out as if seen vertically, the mountain as if from an oblique or lateral view. The image provides us with a view of how the town was organized and how it looked, what its location was, how it was situated in the landscape, and something about how it related to local resources. Further, the view has been compressed to bring the mountain much closer in relationship to the town, which seems to emphasize the contrast between the grid of human control with the unpredictably violent behavior of nature. So this is more than a map, but a highly mediated and thus expressive aerial view of the world.

There are typically two classes into which aerial images fall, the vertical and the oblique. The first is made looking straight down, the second at an angle to the ground. The first tends to appear flat, to reveal pattern, and to be more map-like, while the second is more three-dimensional and pictorial, and establishes context and relationships among features. The Çatalhöyük image does both. Petroglyphs and pictographs—images respectively pecked into or painted on rock, which we tend to presumptively label "rock art"— include representations of landscape features from as early as forty thousand years ago. Elements of terrain in plan, such as rivers and the placement of habitats, appear in rock art as early as the Upper Paleolithic; and by the Bronze Age in Europe simple maps in plan appear to have been widespread. But the Çatalhöyük mural is the earliest image we've uncovered that incorporates both landscape and map in one composition.

It's as if the muralists had not only understood where they were living and how to represent it, but also then discovered a way to fold together a map and a landscape painting to form a stage set. The architecture of the town was blocked out on the floor and the scenery was hung behind it like a backdrop. While researching the historical intersections of art and map, I wouldn't find images constructed like this again until the coastal profiles drawn by Portuguese navigators in the fifteenth century. Their combination of plan and elevation was a technology so useful in depicting coastlines that the Dutch sailors adopted it in the next century. The British Navy subsequently hired Dutch artists to teach its officers how to paint coastal profiles as they sailed around the globe in the seventeenth century. Picturing a space is not just part of creating place, but of claiming it, of converting the terrain into territory.

The sixteen-foot-long fragment of the Anatolian mural may have been part of an image that encircled all four walls of the room, a room that first appeared to James Mellaart to have been a shrine. His predecessors have since determined that most, if not all, of the spaces Mellaart designated as shrines were, in fact, ritualistically decorated rooms in houses. Although it's possible that the mural was meant to be a permanent image, some of the interior walls excavated in Çatalhöyük have up to 450 separate layers of plaster, and it seems more likely it was destined to be plastered over.

I had so many questions about this artifact. Not only did I not understand how its creator(s) imagined its perspective, I wondered why he or she or they made it—for worship, aesthetic pleasure, urban planning? There's evidence that the latter, by the way, was practiced at Çatalhöyük—house designs and brick sizes were standardized. Or did it serve more than one purpose? And why was it the only such image recovered in a site that has offered up many murals, but no others that are depictions of landscape? Would others be found as the site was excavated over subsequent decades? And, finally, what effect on the people living there did such an image produce? All that was clear to me is that the inhabitants of Çatalhöyük had deployed a sophisticated aerial imagination.

The second picture I was pondering, one that suggested an opposition to the "volcano painting," was a photograph made in 2001 by the San Francisco artist Michael Light. A large black-and-white aerial that was part of an exhibition titled "Some Dry Space," it showed—well, I wasn't sure what at first (Plate 1). Most of the other photographs in the series, all made in the Great Basin and Mojave deserts of Nevada and California, contained clear referents. Low oblique views had been shot across the desert floor to a slanted horizon, which was usually defined by mountains and their shadows in the late afternoon. The deliberately skewed horizon lines were vertiginous, and the sense of flying through a volume of space, versus merely cruising above a sheet of it, was palpable. Railway lines and freight trains, roads and the hard edges of agriculture were cultural clues indicating the size of the landscape and our scale in it. But Michael Light's image was an almost vertical view that, although it displayed elevation change, I couldn't read at first as to whether I was looking down into a valley or at the top of a mountain projecting up.

Numerous objects were scattered on the ground to left and right in the image, but were they cattle, people, trees? Their shadows made them seem as if they were in motion, but were they going up or down? Finally, at the upper portion of the photograph, a black area seemed to indicate I was peering at a summit casting a shadow, and I began to read the landscape as a mountain with trees climbing up it. But, then, why was the ground gray in between the trees on the left and white between those on the right? Ah-ha, I said out loud, snow! Given the geographical context of the other photographs, I guessed that I was looking at Bristlecone pines in the White Mountains, a range that defines the eastern border of the Owens Valley in California. And that's what the image showed, groves of the ancient trees on the mountainside pictured from a thousand feet above the ground at 6:30 P.M.

The photograph violated all the rules of what conventional aerial imagery since Çatalhöyük was supposed to convey. It deliberately made it more difficult to read place, instead of easier. The photo forced us away from place and back toward space, away from landscape and toward land, away from known territory toward unknown terrain. It didn't increase our ability to navigate the space, but re-complicated it, made it mysterious. Technically, the photograph was a purely representational object, but what could have been an aerial landscape documentation, a topographical image, made us contemplate it instead as an abstract object in and of itself, which is to say as an artwork instead of as a map.

Some photographers have all along done that, of course, precisely because they want us to regard the outcome of a transformational process as aesthetic, versus merely a transference of what is being represented. Edward Weston, for example, photographed vertically down onto the waves of a Pacific shoreline and eliminated external references to scale, thus creating lace-like abstracts. Eliot Porter, the great colorist, eliminated the horizon and focused on nature at arm's length, thus emphasizing pattern. And although aerial photographers have long delighted in the intricate geometry of features such as plowed fields, braided river channels, and sand dunes, I couldn't recall having been so dislocated before by a landscape image. Part of my response, I realized, was because the aerial view has been normative since at least the 1950s, when commercial air travel began to become affordable for the majority of people. The daily news and traffic images from helicopters, the ubiquity of aerial images in movies and advertisements—all of those were meant to clarify and contextualize, not confound. Light was obviously up to something else. What did he have in mind? Was aereality fundamental to him as a person, as an artist?

And then there was the story I was first told by Sharon Willoughby. We had met at a conference in Canberra about desert gardens, where I had just given a talk about how grids are used by humans to overcome cognitive dissonance in isotropic environments (that is, how we use regular units of measurements in maps and architecture to allow us to navigate in spaces that appear the same in all directions). The wall mural from Çatalhöyük was the first image I had shown. Afterwards, Willoughby came up to me and related a story about Australian scientist Jim Bowler that went something like this:

There's this archeologist in Melbourne who told me that it takes him about two years to teach his undergraduates to really read an aerial photo, to understand what they're seeing. But he took those same photos out into the bush and showed them to some of the central desert Indigenous people, and they understood them almost immediately, despite the fact they'd never been in an airplane, much less seen aerial photographs before.

She said that she'd recently flown into Alice Springs, and had been look-ing at the landscape below while thinking about what Bowler had told her. Here's what she said in a later e-mail about it:

> I was struck by the fractal patterns in the aerial desert landscape, as the plane flew in closer each landscape pattern seemed to resolve into smaller dots, dry river meanders, into branches, into a single complex path. I began to wonder if the fractal nature of desert landscapes gives people navigating on the ground plan clues to the nature of the whole landscape—therefore making it easier for desert people to understand the aerial plan with benefit of aeroplanes? The artwork of central Australian Indigenous people seems to reflect this kind of understanding.

That is, Sharon proposed that Aborigines, by observing patterns in the desert, were thus enabled to more easily read aerial photos. What she was talking about were the traditional Aboriginal dot paintings, which initially had been executed in sand, but in the mid–twentieth century became the basis for a thriving contemporary art practice in Australia. Here's what Jim Bowler had to add in an e-mail after I had flown back to Los Angeles:

> On the question raised by Sharon, the almost ubiquitous fractal imagery in desert landscapes is amazing. Remarkably this is captured in the paintings of the older men whose perceptions of land are as from an aerial view. The nomads of course had to at least *imagine* what lay over the horizon. Their representation of land in form and colour reflects that perception. I suspect the genuine nature of that perception is now dying out with the last of the truly desert dweller although later artists will no doubt continue the style, albeit without the personal experience.

My understanding of dot paintings was that they weren't representations of the land, but somehow the land itself—and that, in fact, Aborigines didn't dis-tinguish between land and landscape. It was all just land. So how, I wondered, would they view colonial European landscape paintings of Australia from the nineteenth century, or more expressionistic ones done by Australian artists such as Sidney Nolan and Fred Williams in the twentieth century, both of who used aerial perspectives? How, I wondered, had flying in airplanes and looking at aerial photographs influenced contemporary Aborigine artists in their attitudes toward land and how they represented it? And what could our culture, where the aerial view of reality is now fundamental to how we shape land-use policies, among other things—what could we learn from their perceptions?

These questions were of keen interest to me, as I felt that writing about aereality from a solely European and North American standpoint would be a very limited look at what seemed to be a universally shared human ability and desire to form aerial views. When my hosts in Canberra suggested that I should return to investigate all of this more thoroughly, I happily agreed. Australia and the American West were explored at about the same time, and both had developed highly urbanized areas with much of the intervening drylands left empty. Both regions had suffered from expansionist policies, extractive industries, environmental degradation, the displacement of indigenous peoples, and from all these factors a severe reordering of their landscapes. Both arid lands were difficult to navigate, thus had excellent sources of relevant artifacts for me to interrogate, offering compact yet complex lodes of paintings, photographs, and other visual traces. Some of the parallels had already been explored in a 1998 bi-national exhibition, *New Worlds from Old*, that pointed out the similarities of their landscape art in the nineteenth century, but much was left to be done.

At the same time, the land itself and the human interaction with the two lands were different. Australia is a much older continent than North America, its geomorphology infinitely more eroded and oxidized. The indigenous people of Australia are among the oldest on the planet, and their cultural traditions far predate our longest continuously practiced ones. In Australia there would be new information, which the anthropologist Gregory Bateson defined quite correctly as "news of difference." Contrast and compare would have been the way a traditional art historian might have put it.

As I began to frame up a work plan, Mike Light offered to let me accompany him as he flew around the American West photographing its deserts and cities. That, I decided, would be where I would begin, in familiar territory and with contemporary work done above ground that I had traversed on foot and in wheeled vehicles for years. I would have the ability to compare aerial and ground views and develop some awareness of the differences and distances between them, but also how one could inform the other.

Flying with Mike would start in Wendover, a small town on the Nevada-Utah border and on the edge of the Bonneville Salt Flats. It is as surreal a landscape as exists in the United States, as splendidly level and visually empty a field of view as one could desire. In aereality, art and the military have always been closely linked and in tension with one another and Wendover was home during World War II to the largest military reserve in the world, a bombing and gunnery training range that covered thousands of square miles. The military is still very much there, the airspace around Wendover part of a vast electronic aerial warfare range that invisibly covers much of the western United States. But while the military was busy target practicing over the desert in the 1960s and '70s, artists were also flying over it, conducting their own kind of reconnaissance, looking for places to site earthworks. The large land projects of that time were not only sited from the air, but also often known to an audience mostly through aerial photographs, which in turn had influenced their designs. At times it was even difficult to distinguish from the air an earthwork from a military structure.

Flying with Mike would also take us over Los Angeles, a city seventy miles wide, a horizontal entity that our friend the UCLA geographer Denis Cosgrove insisted could only be truly understood from the air. Mike and Denis and I would fly over Lakewood, an early tract-housing suburb often called the West-Coast Levittown. It had been photographed from the air by William Garnett in 1950, images that ever since have been used to condemn sprawl. We would have the opportunity to rephotograph Lakewood from the air, as well as examine the steep social gradients between the now mature residential area and the new developments being erected around Long Beach and Los Angeles, the largest harbor complex in North America. Los Angeles has long been pictured from the air, beginning with imaginary bird's-eye views in the nineteenth century through the nighttime grids of the painters Ed Ruscha and Peter Alexander, and the contemporary photography of David Maisel, Mike, and others. How the three media related to each other, the maps to paintings to photographs, would be a valuable nexus toward understanding the cognitive issues at hand.

Then I received an opportunity to work as a visiting scholar for a summer near the Hudson River Valley at the Clark Art Institute, which would occur after flying with Mike and before going to Australia. I began to sense a possible trilogy of essays set "out west," then "back east," and finally "down under," terms which would encourage both writer and reader to think about how we position ourselves in relation to territory. If written in that order, I would be going forward in time but simultaneously back in history. I would move from the contemporary Western United States to the historical East, and then to the much more ancient Terra Australis.

The Hudson River Valley is where the United States developed its first national art, painters such as Thomas Cole and Frederic Church visually reinforcing the literary equation of wilderness with nature with paradise. This was an idea that undergirded Manifest Destiny, with its dual attitudes of extraction from and conservation of land. Flying around the Hudson Valley and its art history would be a way to probe our national identity, and understand how the elevated views used by its painters shaped how we have since perceived and treated the land. This was also prime cultural territory that had been investigated in the *New Worlds From Old* exhibition, thus a bridge to my Australian work. And I could counterbalance the Edenic, prelapsarian view of America developed in the Hudson Valley by flying with Matthew Coolidge from the Center for Land Use Interpretation (CLUI) to investigate the very much lapsed site for an earthworks sculpture atop a landfill in New Jersey.

Although I was pleased with an opportunity to triangulate my topic from three distinctly different, yet resonant geographies, I had a concern I couldn't shake. How would it be possible to talk about flying in the proximity of New York City without dealing with the events of September 11, 2001? *Aereality* had entirely different meanings before and after that day. When Joe Thompson, the director of the Massachusetts Museum of Contemporary Art, suggested that we actually fly the route taken by one of the hijacked planes down the river and into the city, I knew I had an extraordinary opportunity to investigate a concatenation of art and history, as well as that of opposing aerial ideologies: one represented by the minaret and the other by the skyscraper. The flight would also allow me to contrast the depiction of the essentially horizontal nature of Los Angeles with the verticality of New York. Alfred Stieglitz and Georgia O'Keeffe were making views out of their apartment windows early in the twentieth century, a practice growing out of both European painting and photography, and continued by artists such as the New York aerial painter Yvonne Jacquette, who had in the late 1990s been an artist-in-residence with the World Views residency program on the 107th floor of the World Trade Center.

While arranging logistics for fieldwork, I was also searching libraries and museums for relevant literature. Although there was a plethora of cof-feetable books featuring views made "Above . . ." almost every country and major city in the world, the only overview of aerial images that I could find was a seventy-two-page exhibition catalog from 1983 titled *The View from Above*. The introduction was by Beaumont Newhall, photograher and first curator of photography at the Museum of Modern Art, who had worked in aerial photo reconnaissance during World War II. He started the essay by saying how surprised he was by the lack of attention accorded the aerial view by artists and art critics. Although the exhibition (curated by Rupert Martin for The Photographers Club in London) was limited to a time span of only 125 years, it nonetheless captured the essential evolution of aerial images from those made in balloons through ones compiled via satel-lite. Despite the paucity of aerial image histories, I found that artists have been fascinated by elevated views all along, taking their sketchbooks, easels, and cameras to the highest point of every vantage they could find. That was true of Dutch panoramic painters in the 1600s, the Hudson River painters during the 1800s, and contemporary artists such as Gerhard Richter and Ed Ruscha. With the introduction of internet tools such as Google Earth, aereality has become a space we take for granted. But if Newhall were writing his introduction today, I suspect he would start it the same way. Many critics and curators still have trouble accepting aerial photographs as aesthetic objects.

This is a situation familiar to me, given that I write about art based on landscape, be it paintings, photographs, sculptures, performances, or instal-lations. Even before Joshua Reynolds stuck his nose up in the air over Dutch landscape art and declared it to be merely "map work," curators have strug-gled with how to value the representation of terrain with other kinds of images, such as history paintings and portraits, or what at times has been the even more privileged discourse of abstraction. Aerial photographs, by far the majority of which are produced for utilitarian purposes, are likewise more often seen in cartographic terms than artistic ones.

My reaction is that aerial images are now more important than ever to study as aesthetic objects within the broader range of visual culture, precisely because they are so widespread in their use and influence. Aerial representations of the earth's surface, whether photographic or remotely sensed and digitally reconstructed images, are used for everything from land-use planning to military campaigns. This has been true since at least Leonardo's 1502 views of the River Arno done during his stint as the chief engineer for Cesare Borgia. His drawings were as much for strategic purposes as aesthetic ones, despite how we read them today as part of his artistic oeuvre. Furthermore, aerial imagery is now infinitely penetrable; you can interpolate any kind of digital information into it and map any kind of change upon it. It is perhaps at this point of malleability when aesthetics actually provide more efficient and relevant algorithms for parsing images than mathematics.

It became my conviction that more than ever we needed to understand why and how we accord such authority to views elevated literally and figuratively above all others. It seemed valuable to apply to this entire class of images the same questions I was asking of the Çatalhöyük mural, Mike's photograph of the bristlecone groves, and Jim Bowler's story about Aborigines. And it was definitely time to go flying. If our view of the world has been increasingly aerialized since World War II—in part because our view of that war was so dominated by aerial imagery—flying out of a former military base from that era would "ground" my writing in a specific set of circumstances.

# 1

# OUT WEST

# 1. Targeting

*Trying to fly without being prepared to say it's impossible is like trying to learn to swim without water, although we must admit that swimming can be flying in water.*

**—Peter Greenaway**

**The tiny two-seater aircraft seems to walk down the runway, break** into a run, and then step up into the air with a small wobble. Michael Light pulls back on the stick and we rise quickly into the morning air over the tarmac. This is the way the dream of flying begins: gathering yourself for a jump, making a mental leap, and then gliding above the ground. Among all the reasons for flying—travel, recreation, reconnaissance or surveillance to name just a few—the re-creation of a dream may be among the most potent. Light's father, a bombardier during World War II, was blown out of an airplane during a mission over France, and the photographer has been blessed or plagued with dreams of that event since childhood. Plagued in that the dreams are not always comfortable, sometimes downright terrifying. But blessed because they have provided him with an inexhaustible muse. His father survived, by the way, his parachute opening just in time to prevent his death.

This mid-April morning we are flying out of Wendover, a small town on the Nevada-Utah border where Interstate 80 debouches from the mountains to the west and enters the longest stretch of straight pavement in the entire national system of freeways. The road traverses the Bonneville

Salt Flats without so much as a single exit for thirty-seven miles to mar the purity of the pavement. Signs instruct drivers to pull off the freeway and take a nap on the shoulder should they feel themselves falling asleep, lulled by the incessant and undeviating whine of tires mile after mile. The freeway also marks the only corridor across this part of the Great Basin Desert where one can fly without straying into military airspace, a passageway four miles wide on both sides of the pavement.

In addition to the gamblers driving in from Utah, the automotive enthusiasts chasing land speed records out on the salt flats, and a variety of military personnel in town for training exercises, Wendover also hosts a steady stream of artists and writers. The Center for Land Use Interpretation (CLUI) established its first office during the mid-1990s in Los Angeles, but now runs several facilities across the United States devoted to documenting and understanding how we perceive, carve up, and attempt to control terrain. The nonprofit organization relies on a volunteer corps of geographers, anthropologists and others trained in a variety of sciences and humanities to conduct the work, but the organization also involves artists as key investigators.

Wendover is a site where humans have been the primary geomorphological agent since the early twentieth century, which is to say we've shaped the land here more than rain, which used to be the primary sculptor of the desert surface. Evaporation ponds built for the extraction of salts and other minerals are spread out over hundreds of square miles to the east, and to the south are thousands of square miles held by the military that have been dug into, blasted, bombed, and tracked over since World War II. A few dozen of the more than six hundred buildings from the war still remain, and CLUI rents several of them for exhibition spaces and workshops. In 1996 CLUI hauled a single-wide up from L.A. as a residency unit and placed it on one of the old airbase foundations.

The artists and writers who apply from around the world to work here are usually treading a line that wanders into the practice of geography, and I'm no exception. I've been working here on and off for five years with the photographer Mark Klett on a book about the *Enola Gay* and the erosion of history; and for the last two years on an experimental cartographic exploration with the artist Katherine Bash. Despite the fact that both projects involve the use of aerial imagery, I've never flown out of Wendover to explore the local territory. Neither had Mike, which is why he has joined me for a few days of flying here. Both Mark and Katherine have also expressed their desire to look at Wendover from above, curious to see how the elevated vantage would enlarge their experience of the desert. And yes, I feel a twinge of guilt to be up here with Mike looking down on the airfield.

As the architecture critic Rayner Banham pointed out in 1962, the development of both commercial aircraft and airports started out as a function of military surplus. It's difficult to pin down what was the first airport, but England's Croydon is often cited. Before World War I, airfields (also called aerodromes) were merely grassy fields available for the launching and landing of aircraft, but by the time the war ended actual runways had been constructed, hangars built, and facilities for pilots and passengers included. The armistice with Germany was signed in November 1918; in February of the next year commercial air traffic was established between Berlin and Weimar. That year saw the first large aircraft for commercial flights put into service, ten converted Handley-Page bombers, biplanes reconfigured to carry twelve passengers apiece on the London-to-Paris route. Two adjacent airfields built in 1915 and 1918 to protect London from Zeppelin attacks were linked and opened as the Croydon Aerodrome in early 1920. The evolution of both flight and photography were force-fed by various militaries.

Just two decades after the opening of Croydon, the Wendover Air Field was on its way to becoming the largest military reserve in the world, a bombing and gunnery range where pilots learned the skills they needed to firebomb Hitler's Third Reich into surrender. It's also where the *Enola Gay* was hangared in between the Boeing factory in Ohio and flying off into the Pacific to drop the world's first atomic bomb on Hiroshima. Its enormous hangar, which stands across the tarmac from the CLUI unit, is now below us and to the left of our position overhead. From above it's obvious how its rounded roof of corrugated metal is slowly being flayed to pieces by the daily desert winds.

The World War II bombers that flew out of Wendover included the largest, heaviest, and most expensive war machine ever made at the time, the B-29 Superfortress. Its wings were as long as a football field and it weighed in at more than 100,000 pounds. Mike's brand-new "CTSW" made by Flight Design of Germany is as far down the size scale as you can go and still be classified by the Federal Aviation Administration (FAA) as an airplane. Flight Design started out in the 1980s as a builder of ultralights, which I always think of as model airplanes barely large enough to carry a single person. According to FAA regulations, ultralights aren't allowed to carry a passenger, fly over towns, fly at night or in clouds, or weigh more than 254 pounds. The Light Sport Aircraft category was designated in 2004 by the agency as those aircraft occupying the niche in between ultralights and general aviation aircraft, such as those larger two- and four-seater Cessnas and Beechcraft that we're used to seeing at small airports.

Mike's plane, which has been compared to a flying egg, is a bulbous yet sleek pod with a wing that all together weighs about 650 pounds, has a wingspan of 28 feet, and is made of carbon fiber and Kevlar, the material used in bulletproof vests. Its wing is cantilevered over the fuselage without struts, which makes it an ideal photographic platform. It doesn't hurt that it has a phenomenally low stall speed, 39 knots, or about 45 miles per hour. That's why it feels as if we're walking. You can almost park it in one spot over the ground while you work, making the experience more like flying in a helicopter than your average fixed-wing aircraft. Sitting in its molded seats bolted to the floor, your feet stuck straight out ahead of you, is a posture akin to riding in a "bathtub" Porsche sports car from the 1960s. It's spartan, yet comfortable. As an added virtue, there's a red handle in between and slightly behind us. Yank on it in an emergency, and a parachute rockets out of the wing overhead. The entire aircraft floats down to earth. What's not to like?

Mike puts us on a course heading east-northeast, toward Floating Island, a peak that rises nine hundred feet above the salt flats and marks the end of the Bonneville Race Track. It "floats" because it is sometimes surrounded by water, but most often just appears that way due to the prevailing mirages created by the hot air hovering just above the even hotter ground. The difference in temperature produces a thermal discontinuity, a boundary that acts as a mirror line. The visual distortions here contributed to the crashes suffered by pilots in training during the war, and remain a subject of frequent interests for visiting artists. From the air, however, it has no float; it's obviously and solidly a part of the ground.

We cross the interstate at about five hundred feet and are almost immediately over the Intrepid Potash evaporation ponds, a series of bright blue and green rectangles that look like they are holding captive parts of the Caribbean. Various sodium and potash products have been mined here since the early twentieth century, and some of the ponds are more than a square mile in size. Workers pump dyes into the ponds to darken the water and hasten natural evaporation, hence the brilliant hues. Once you know it's not an organic process, it looks suspiciously toxic. This isn't table salt that's being produced, however, but the magnesium chloride we spread on roads to suppress dust, and potash for agricultural fertilizer.

The ponds tend to be rectilinear, and although the ground here is almost flat, they are arranged in an enormous semicircle that follows the subtle contour lines of an ancient lake. Humans tend to organize land into various geometries, part of the way we transform land into landscape, and a process that is more apparent in arid lands than elsewhere with little vegetation to hide the evidence. Once past the ponds, another kind of patterning on the ground becomes visible. When flat ground dries, it shrinks, which causes stress. Nature seeks to distribute stress as evenly as it can, and the physics of mud and salt flats dictate that as they dry the ground breaks up into polygons. They range from triangles through five- and six-sided figures. Look at a mud puddle after a rainstorm and you'll see the same thing—or on ground over permafrost, the surface of frozen lakes, and the geodesic domes of Buckminster Fuller. It's all the same mathematics.

Photographers working at CLUI often make images of the patterned ground, a trope for the entropy so visible in the desert. What's uniquely apparent from the air is how fractal the pattern becomes. The smallest polygons are the size of your hand. When you're walking across the salt flats, you can see how they organize into larger sets of the same kinds of polygons the size of a tabletop. From the air it's obvious that those, in turn, form even larger patterns many square yards in extent, and those aggregate into still more massive figures. It's a fractal phenomenon because it is self-similar at all scales, and it reinforces the cracking of the earth's skin as a metaphor of time winding down in an entropic heat-death.

We follow the raceway straight for Floating Island until the plane's Global Positioning System (GPS) map shows we're about to hit the border of a Military Operations Area, or MOA. The Utah Test and Training Range (UTTR) is the largest combined restricted ground and airspace in the country. While 2,675 square miles of land is reserved for bombing practice, 19,000 square miles of airspace is restricted. It's the largest contiguous authorized space for supersonic operations available to the United States Air Force, and they run twenty-two thousand training sorties a year in it. We see F-16s and other aircraft flying over Wendover every day and night, sometimes so high we can barely hear them, sometimes buzzing the runways a hundred feet overhead. Mike turns around and we follow the interstate back toward town.

Perhaps the next most powerful motivation underlying the human desire for flight, after recreating the dream of it, is to see what's over the next hill, to gain an elevated viewpoint from which to survey one's surroundings. That's our primary agenda this morning, so we turn north over the town and aim for a pass through the hills, the first leg of a clockwise rectangle. We'll fly north to the *Sun Tunnels* sculpture by Nancy Holt, east to the Great Salt Lake, then bear south along the shoreline and across the lake from *Spiral Jetty*, that metaphor of entropy bulldozed out into the lake by Holt's deceased husband, Robert Smithson. Our last leg will head west back to Wendover, the fourth side in our boxing the Great Salt Lake Desert.

The most obvious feature of Wendover from the air is the enormous X formed by the runways that were built in 1942 as ten thousand-foot-long concrete strips eighteen inches above grade—just high enough to avoid flooding during the largest imaginable spring storms. They were one of the deciding factors for Lt. Col. Paul Tibbets when he was scouting a location from which to train his crews in the fifteen specially "silver-plated" B-29s forming the 509th Composite Group, the squadron that was designated the atomic bomb group. The other major factor was Wendover's isolation, which is stunning. The town sits 125 miles west of Salt Lake

City, the nearest urban area, and although it is linked to the rest of the world by the freeway, when the soldiers of the Army Air Corps first arrived in 1940, they thought they had landed on the moon.

The town sits at 4,300 feet above sea level on an ancient bench of sediments, formerly a beach on one of two Pleistocene lakes that covered most of Nevada and Utah. The lakes were contained by an anomaly on the North American continent—a series of contiguous and enclosed watersheds with no outlet to the sea. We know this 200,000-square-mile region as the Great Basin, the largest, highest, and coldest of the five major deserts in the country. The body of water covering Utah, and the future site of Wendover, was Lake Bonneville. Before it breached its northern shoreline about 16,800 years ago, the lake was 325 miles long, 135 miles wide, and more than a 1,000 feet deep. It covered 20,000 square miles and was the size of Lake Michigan. As the Pleistocene Era ended ten thousand years ago and the lake dried out, it left behind a series of fifty or so separate shorelines and terraces carved by wave action, many of which are visible around Wendover.

The largest remnant of the ancient inland sea is the Great Salt Lake, which is currently about 75 by 35 miles in extent, covers 2,400 square miles, and is 33 feet deep at its maximum. To its west is the Great Salt Lake Desert, the largest playa in our country. The intermittently dry lake bed runs west from the lake to Nevada and covers 4,000 square miles, most of which is labeled with military acronyms on flight charts. On the westernmost edge of that sub-desert are the Bonneville Salt Flats, 400 square miles of blindingly white deposits left behind by the ancient lake and then augmented by erosion from the surrounding mountains. The flats are so hard and, when scraped, so flat that they were used during most of the last century to set world land speed records for cars, motorcycles, and anything else on wheels.

It was Lt. John C. Frémont of the U.S. Army Corps of Topographical Engineers who ascertained in the 1840s that the region, the last geophysical province of the country to be defined, was a vast interior desert with no outlet to the sea. He determined this by following the example of his hero, the greatest naturalist of the nineteenth century, Alexander von Humboldt, who climbed up everything he got near in order to observe unities in the surrounding terrain. Frémont, much to the disgust of his gloomy topographer, a German named Charles Preuss, dragged his companion with the requisite barometer up the front range of the Wasatch to the east of the basin, then months later up the Sierra to the west. Using the barometer they measured hundreds of elevations in order to establish approximate contours, all the while circumnavigating much of what would become Utah and Nevada. Once they determined that no river emptied the Great Salt Lake or any other of the interior depressions, Frémont proclaimed it to be the Great Basin. Humboldt, who was in correspondence with Frémont, was delighted, perhaps not least of all because the young explorer named several features in the basin after the man who inspired an entire century of geographers.

Frémont forged a route west twenty-five miles to the north of Wendover in 1845, and the maps prepared by Preuss guided emigrants through the area starting the very next year. Among them were the members of the ill-fated Donner Party, whose struggles crossing the mud and salt flats delayed them just enough to cause them to be caught in the High Sierra by winter snowstorms. The Pony Express, Overland Stage, and Lincoln Highway—American's first transcontinental motorway—passed sixty miles to the south to avoid the salt flats. In 1907 the Western Pacific Railroad, seeking the most level route to California, established a line that ran straight across the flats. Wendover served as a water stop for the locomotives, a roundhouse was built, and a road soon paralleled the tracks. In 1914 the first transcontinental telephone line linking the two coasts was connected in Wendover, and ten years later a stranded traveler opened a local service station. Gambling was legalized in Nevada in 1931, and the following year Bill Smith installed a roulette table, making his place the town's first gambling establishment, soon to be called the Stateline Cafe.

In January of 1939 Hitler had invaded Poland and war in Europe was beginning to look like an inevitability. The Japanese had been militarizing for decades, were deeply involved in their second invasion of China, and were now making noises about the Pacific Ocean. Then there was the letter that Albert Einstein sent President Franklin D. Roosevelt that summer about the research the Germans were conducting into a bomb based on nuclear fission. Roosevelt correctly read the threats and asked Congress

for funds to expand the Army Air Corps. Several new bases were established in the arid West where the flying weather was good most of the
year, the land wide open for bombing practice, and the facilities far from
any possible enemy attack. The first buildings on the Wendover Air Field
base, constructed of wood and tarpaper, were meant to be only temporary
structures. When the airmen arrived in September of 1940 to begin construction on the base, the town had a population of 103, its hotel and service station, a drugstore, and a grocery.

At the height of the war, more than twenty thousand personnel were
stationed in the base's 663 buildings, many of which are still standing
today, albeit decaying as time passes and memories erode. The base was
decommissioned after the war, and although the military still comes to
town during exercises, mostly the airport is used for general aviation—
private aircraft—and a daily Boeing 737 that brings gamblers in from
Canada. A variety of experimental aircraft also visit the airport on a regular basis, but even so, Mike's plane has elicited more than a few stares.


We're now passing over the base and what has become a strangely bifurcated town, caused by the fact that gambling is legal in Nevada and not
in Utah. In the 1990s corporate gaming came to Wendover with the erec-
tion of the first of several highrise hotel-casinos. The population of the
original town in Utah is stable at about 2,800 people, mostly Hispanic
workers employed on the Nevada side, where the population is more
than twice that size and booming. The socioeconomic gradient is obvious
from the air. In Utah the town is mostly abandoned barracks and newer
mobile homes surrounded by brown yards featuring sagebrush and tum-
bleweeds. In Nevada there are well-watered lawns, a golf course, and
most of the retail businesses.

I look down at the crossed concrete strips. At night during the war the
bombers returning from exercises to the east would fly in low over the salt
flats, which were sometimes covered in a few inches of water. The temporary lake could be so still that the Milky Way would be reflected almost perfectly in it, and the pilots didn't know if they were flying rightside up or
upside down. It was always a relief when the brightly lit X of the Wendover

runways came into view, although not a guarantee of a safe landing. More than one of the cumbersome B-29s plowed nose first into the playa while attempting to land.

Mike Light's father, Robert Taugner, died when his son was eighteen, which made his great-uncle Dr. Richard Light (a well-known aviator of the early twentieth century) the boy's effective grandfather. Robert was the bombardier and navigator aboard a medium-sized bomber flying out of England six days before the D-Day invasion on a raid over a marshaling yard in France when they encountered heavy flak. He just had time to open an escape hatch when the plane was hit and the explosion blew him out of the opening. He awoke to find himself falling through the air without having deployed his parachute. Although he managed to open the chute in time, he impaled himself on a fence. Rescued by the French, captured by the Germans, put into a camp—he survived, but the event not only scarred his body, it also indelibly marked the mind of his son. This morning's flight is marked by periods of silence as the artist/pilot absorbs the resurfacing memories of his father.

Mike's first significant artwork dates from 1989 when he was in art school: he turned his father's wartime narrative into a series of postcards, as if his parent were writing to him from beyond the grave, and illustrated them with found photos. In 1994 he exhibited his San Francisco Art Institute thesis project, "Blue Fall," and he describes it this way:

> For my MFA show I exhibited about thirty-five black-and-white aerials that I'd made on commercial flights from twenty-five to thirty thousand feet, toned blue and then encased in steel boxes set about eight inches off the floor. Very cold, very industrial. 16 x 20 inch images with glass on top, and frosted letters mounted inside the glass with the names of American inventors, some famous like Edison and Bell, but mostly obscure ones, like the guy who invented the microwave. So viewers walked through this landscape and looked down on it and the names weren't visible from directly above, but from an oblique angle they became legible, a conflation of our country with industry.

Above it all was a TV monitor floating fifteen feet up, a disembodied eye looking down. I appropriated some looping clips, one of a man jumping out of a plane and filmed from above, his parachute never opening but implied, the ground spinning around and around. You craned your head back and looked at a man falling upwards. The blue of the television was resonant with the blue of the photos, and there was also a clip of a military man's finger pushing a World War II–era button over and over.

Mike's first book *Ranch*, published the following year, is a kind of "photo novel" made on one of the last large cattle ranches in central California, and it starts with aerial photos that descend in sequence as if we are falling to the land. Mike never flies without that vertiginous memory somewhere in mind, the tension between flying and falling a creative *frisson* for him.

It's obviously poignant for him to fly over the old air base, given his father's experience in the war, but flying and photography have long been linked in the artist's family. In 1934–1935 his great-uncle Dr. Richard Light, who had learned how to fly in the U.S. Army Air Corps, became the first person to fly around the world in a private seaplane, a feat accomplished only seven years after Lindbergh first crossed the Atlantic. In 1937–1938 Richard took to the air on an extended journey with his new bride, Mary Upjohn, who would serve as copilot, navigator, and photographer. They flew south in their Bellanca Skyrocket, a six-seat monoplane built in Delaware with a wingspan of fifty feet, first from Michigan to Texas, then down through Central America and the west coast of South America, and finally crossing the Andes over to Rio de Janeiro. The plane had a roll-down window on the passenger side, and Mary hung out of it with a 5x7 Fairchild camera. Clutching the heavy camera as Richard banked the plane to provide her with a clear view was a continuing test of her strength, but she took some remarkable shots.

It's somewhat astonishing to me that the U.S. Navy Photographer Lt. George R. Johnson had in 1931 somehow missed what are now the most famous geoglpyhs in the world, the Nazca Lines, but it was Mary who first photographed them, albeit unwittingly, in October 1937. The lines, which cover almost 200 square miles and include geometric and zoomorphic figures more than a thousand feet in length, were made by native peoples sweeping aside the highly oxidized rocks to expose the lighter ground underneath. The figures were first mentioned by Europeans in 1547, when the Spanish explorer Pedro de Cieza de León described some of the lines visible on a hillside, and have remained a controversial source of speculation for anthropologists ever since 1927, when a Peruvian doctor began to study them. Their essential mystery was how people could have designed and then executed such large figures without an aerial vantage point. Even today some people insist on interpreting them as landing instructions for alien spacecraft.

After this first six-week trip for the newlyweds, they shipped their plane from Rio to Cape Town, then proceeded to fly a twenty-thousand mile photo transect up the entire east coast of Africa to Cairo, and the images Mary Upjohn took over Africa remain the only record of the continent from the air during that time. The book they published, *Focus on Africa*, contains 323 photos, almost all of them aerials. All are arranged in regional groupings in chronological order and captioned by Richard. The text of the book is more than a straightforward account of the trip, however. By turns dramatic and humorous, it includes historical introspection and rudimentary socioeconomic analysis, and is overall a surprisingly good read for a book by a neurosurgeon acting as an aerial geographer.

Richard asks a key question. They're preparing to fly across Basutoland and Swaziland, two countries in which they will not so much as set down to refuel, when he asks: "Can a person visit a country yet never set foot upon it? Does an airplane journey across a territory entitle the traveler to claim that 'he has been there'?" Richard has no answer, and quite sensibly says so—then goes about describing that part of the trip. The photographs taken of the continent may not entitle him and Mary to citizenship, but overall they reveal important geophysical information, including the differences between native and colonial patterns of settlement. The circular villages of the former are in stark contrast with the rectilinear grids of the latter, for example, as is a reliance upon irrigation. While the Europeans channeled the water to fit the grid, the Africans planted along the existing watercourses, following and preserving to a greater degree the existing topography.

The clearing of bush by the settlers in order to deny the tsetse fly habitat, while it on the one hand decreased disease, on the other it created erosion problems which from the air were all too evident and prescient. The Europeans had recently introduced contour plowing to alleviate the problem, but also erected animal barriers to prevent the migration of domesticated animals and their insects. Over time that would decrease game while allowing human populations to increase, with better health and agricultural nutrition fueling expansion. For every action a reaction, and from the vantage point of hindsight, disaster awaits in the coming decades. Richard may not have been able to say he'd truly been in Swaziland, but he was learning from it nonetheless. For a lapsed climber such as myself, the most amazing photographs are those taken from as high as 23,000 feet over Mt. Kilimanjaro and peering down into its concentric crater rings, and a wonderful five-picture aerial series stretching from its snowy summit across the saddle to its secondary peak. The panorama of the nearly 2,000-foot-deep Ngoronogoro Crater 125 miles to the west is likewise astounding. Now it is a national park, but the 13-mile-wide depression then was a legendary, almost supernatural sanctuary for African megafauna. To cap it all off, Richard and Mary took off in fog early on a late December morning and broke into a clear sky through which they were able to approach and photograph the most elusive mountain range in the world, what Ptolemy may have been referring to when he named the "Mountains of the Moon" as the source of the Nile. Today we know the range, which sits on the border of the Congo and Uganda, as the Rwenzori. The range remained undiscovered for years because it's hidden in clouds almost year-round, and even today its peaks are seldom climbed because of the difficulty of approach, including intense local corruption and political unrest.

Mary spent most of the time hanging out of the window making panoramas from the east, west, and south, plus portraits of the individual peaks. Richard reported that she was blue from the exposure by noon, and that upon landing she spent the next three days in bed with a fever. Once she recovered, they continued their journey, eventually making it to

Cairo and taking oblique shots westward across the city and the Nile of the pyramids on the other side. Beyond the monuments stretched sand to the horizon. It's one of the finest pictures I've seen from any age that demonstrates how land and landscape meet.

Mike has long had a copy of *Focus on Africa* in his possession. It's a classic of exploration literature, a milestone of aerial photography, and of course it influenced the construction of the project for which he first became widely known, *Full Moon*. In 1994, after noticing how lunar the Nevada desert appeared to him and comparing it with the images he'd seen that had been made by the Apollo astronauts from 1967 to 1972, he approached NASA for a more thorough look. He ended up going through many of the 17,000 negatives taken from a hand-held on the moon and the 15,000 automatic shots made from orbit. He then edited them down to 130 or so images sequenced as if they were from a single flight made from the Earth to the Moon and back again.

If the photos from *Full Moon* help make the strange familiar, then Mike's ongoing series of aerial work makes the familiar once again strange. Since 2000 he's been photographing from small aircraft above the Owens River Valley east of the Sierra Nevada, and from helicopters over Los Angeles. The series, mostly black-and-white aerials shot at an oblique angle to the landscape, disassociates both the desert floor and the city grid from any familiar view, land-based or aerial. Where we usually deploy a number of visual cues to enable us to understand the rural and urban deserts of America, Mike decouples them from frames of reference by shooting into the sun, eliminating clues about size and scale, and making the Mojave Desert as alien as the moon.

After creating several series in this large body of work, Mike decided he had to finally bite the bullet and join the family tradition. Although, as he confesses, "I learned to fly a glider before I knew how to drive a car, soloed when I was fourteen, and got my ticket when I was sixteen," he had never himself owned a plane. It was only late this last year that he bought the German Light Sport craft and was awarded his license for powered flight. His flight from the Bay Area, where he lives, across all of California and Nevada, took four hours and four minutes. That specificity of detail is typical of Mike, as it is for many photographers and pilots. His flight from the

Bay Area to Wendover was his longest to date and he was physically vibrating when he stepped out of the plane and onto the tarmac yesterday.

Today he's taking very few photos of the ground, concentrating instead on the unfamiliar airspace and the necessity to pay attention to the nearby boundaries of the MOAs. Once we're past the town, he negotiates Leppy Pass between the steep-walled limestone mountains to the north. As we go over the ridge we enter the Pilot Basin across which stands Pilot Peak, a ten-thousand-foot mountain that was a landmark critical to Frémont and the subsequent emigrants coming across the desert. Numerous Lake Bonneville shorelines trace huge arcs that sweep around this end of the valley. I've walked through this area several times, and never noticed the ancient beaches before, although presumably a geologist would have noted the lines of gravel underfoot. We spot a small concrete bunker and a bomb crater, and then large but faint arrows and concentric rings graded on the desert floor, what's left of an illegal bombing range from the war. The American military has a habit of stepping outside officially sanctioned bombing ranges, and this target is well to the north and west of historical and contemporary restricted space.

Mike makes a circle so we can look more closely at what was known as the Pilot Peak Target Area. To target something is to express a desire to possess it in some way, and bombing has been a governmentally sanctioned method of gaining purchase over terrain ever since the Mongolfier brothers lofted the first manned balloon in 1783. In Paris that day was Benjamin Franklin, who was helping negotiate the peace treaty between the Americans and British forces. As the balloon went aloft, he noted the potential military applications for reconnaissance and the next year the French tested the first aerostatic reconnaissance balloon. The pilot reported that he could see a town eighteen miles distant through his telescope, a result that so impressed the government that Napoleon promptly established the world's first air force, which proved to be a decisive factor at the Battle of Fleurus that same year.

What the pilots from Wendover did from 1942 through 1945 was fly over a variety of artifacts and attempt to hit them. A full-sized and realistically shaped warship was made of tar out on the salt flats so it would be visible enough to bomb. A city of salt was constructed on the other side of the hills we just flew through. Barrels of oil were set out in a circle and lit on fire at night. All were targets. Some of the bombs, such as the ten thousand concrete-filled bomb dummies the *Enola Gay* crew used for practice, buried themselves thirty-five feet deep in the mud when they hit. The crews trained at Wendover were the best in the world, able to position their silk spider-thread crosshairs over anything on the planet. Despite the fact that we dropped the world's first two atomic bombs, within a few days during late August 1945 incinerating more than 150,000 Japanese, mostly women, children, and the elderly unable to fight, our military does seem to genuinely prefer weapons that minimize collateral damage. Hence the laser- and satellite-guided bombs that we can now fly down chimneys and into office windows.

You cannot fly around the West without having such thoughts, the restricted airspaces are so expansive and obvious. The size of the Nevada Test Site alone equals all of Rhode Island with room to spare. The corridors through which electronic warfare are conducted stretch almost from Mexico to Canada, and you really, really don't want to run into so much as the turbulent air left behind by a military jet traveling faster than the speed of sound if you're in a vehicle weighing a quarter that of a Volkswagen Beetle.

Mike heads west and then north around Pilot Peak. We hope to cut a corner across the restricted air space, which is allowed if you have permission from the local authorities, and he tries repeatedly over the radio to raise Hill Air Force Base north of Salt Lake City, which administers the UTTR. No response. We're either too far away or they're simply not concerned with an aircraft as small and low and slow as ours. Hill AFB is the site where the military services all the country's Intercontinental Ballistic Missiles, but also our military reconnaissance and photo gear. Its runways host a total of eighty-five thousand operations a year, making it the busiest airport in the U.S. military. We keep trying to raise their control, using the frequency on the aviation charts, but no answer, so we skirt the MOA instead of cutting across it and head toward our next target, the large X formed by *Sun Tunnels*, a sculpture by Nancy Holt.

# 2. Earthworks

*Restricted areas contain airspace identified by an area on the surface of the earth within which the flight of aircraft, while not wholly prohibited, is subject to restrictions. Activities within these areas must be confined because of their nature or limitations imposed upon aircraft operations that are not a part of those activities or both. Restricted areas denote the existence of unusual, often invisible, hazards to aircraft such as artillery firing, aerial gunnery, or guided missiles. Penetration of restricted areas without authorization from the using or controlling agency may be extremely hazardous to the aircraft and its occupants.*

—Federal Aviation Authority

**As we round the Pilot Range, I look out my window and to the south** spot Lemay Island, another mountain peak. What I'm seeing is only the few hundred uppermost feet of what is, in fact, a much larger mountain, most of which is buried in several thousand feet of accumulated silts and salts on the flat. Nancy Holt and her husband Robert Smithson bought forty acres on the shoreline of the island in 1969, a wedge of gravel bench overlooking an embayment of the surrounding playa. Smithson had hoped to situate a sculpture there, but was lured away by the notion he could place a work out in the red algal waters of the Great Salt Lake itself. He scouted several places around the lake before discovering the abandoned oil works at Rozel Point across the lake, and in spring 1970 constructed *Spiral Jetty* there, one of the most iconic of all contemporary American art-works. The 1,500-foot-long counterclockwise line of basalt and earth bull-dozed out from the shore and into the lake was a major intervention, not just in the land but also in the art world—a challenge to the accepted norms of how art was created in studios, shown in galleries, and sold to collectors and museums. Along with sculptures and drawings created previously on the land by Michael Heizer and Walter De Maria, *Spiral Jetty* helped define that branch of art we call Earthworks, or Land Art. And that genre has everything to do with the aerial imagination exercised by humans.

Earthworks arose as a genre in the late 1960s. On one level it was made possible by the increasing levels of abstraction that had been exhibited and marketed in New York City since Jackson Pollock and the Abstract Expressionists of the 1940s and '50s. Those visually dense works were fol-lowed by the increasingly pared-down and formal visual fields of Barnett Newman and Mark Rothko, a trend that culminated in the severe serial geometries of minimalism: bricks stacked on the floor by Carl Andre, Sol LeWitts's cubic matrices, and Robert Morris's polyhedra are good exam-ples. All of them used geometrical figures the *gestalt* of which could be apprehended the minute you walked into a gallery. To paraphrase Claude Levi-Strauss, you saw the whole before you perceived the constituent parts. Such forms could easily be magnified up into an architectural scale and placed outside, and soon were.

On a more pragmatic level, as the works became larger and more expansive, working outside was less expensive than dealing with a studio space or gallery. And land in the arid West was cheapest of all. As Michael Heizer has been at pains to make plain to me and everyone else over the years: "Material is place and place is material." His point is that he didn't work in the West to be scenic, or to include the desert as a component of his sculptures. He could just as easily have worked in New York if it hadn't been so expensive and confining. The rock, gravel and sand he needs are cheapest in the desert. The other thing Heizer insists on is that he doesn't work with scale, but with size. And size has always mattered to him. The largest structures in the world, from the pyramids to the Three Gorges Dam to the American interstate system, are all made of earth, be it stone, mud, or concrete.

Heizer, who lives in south central Nevada where he is constructing *City*, the largest single artwork on the planet, was born in 1944 to Robert F. Heizer, a prominent anthropologist who specialized in how early cultures moved large stones. The senior Heizer taught at the University of California, Berkeley, but Michael spent much of his childhood accompanying his father to excavations in Nevada, Mexico, Peru, and Bolivia, along the way experiencing the joys of conducting aerial surveys, digging in the dirt with everything from toothbrushes to bulldozers, and picking up a vocabulary of sculptural shapes from places as varied as Karnak, the ancient burial city of the Egyptians, to La Venta, where the Olmecs constructed ceremonial ball courts.

Michael Heizer's grandparents and uncles were mining engineers in California and Nevada, and his father was an expert in the rock art of the Great Basin. When the young artist grew frustrated with the New York art scene in the mid-1960s, he flew back out to the desert. In 1967 he started creating minimalist geometrical voids in the Sierra Nevada, and the next year on the playas to the east. That year he did what any modern archeologist—or miner—would do, and found a pilot so he could locate likely sites for future investigation. He hired Guido Robert Deiro, who a couple years previously had been flying the secretive Howard Hughes around the desert. The pilot and artist found likely art sites on the playas of the Mojave and Great Basin not only for Heizer, but also for his friend and fellow artist Walter De Maria. Heizer created numerous works out in the desert during 1968. He drew figures with motorcycles leaving tracks on the ground, threw dye into the wind to paint the ground wherever it fell, and dug a series of trenches stretched out over 540 miles, a work collectively known as the *Nine Nevada Depressions*. Some of the component sculptures were fifty feet in length, others more than a hundred. You could experience the excavations from the ground, but they revealed their individual and serial nature more fully from the air. Numerous photographs from that year often include the airplane and helicopter from which Heizer was working with Deiro.

In July of that year Heizer took Robert Smithson and Nancy Holt on a month-long tour of the region, including Mono Lake, a Pleistocene remnant to the north of the Mojave. The three artists made a short film there, and Smithson was taken with the idea of siting a sculpture on the shore. He later decided instead on the Great Salt Lake after he heard about the algae turning its waters red. Smithson was an avid enthusiast of cartography, photo maps made by NASA, and the work of Buckminster Fuller, whose "Dymaxion" map projection and geodesic domes utilized the same mathematics as those underlying the polygonal stress fracturing on desert playas. Smithson, who like Heizer was fascinated with crystalline structures, was well prepared to adapt the strategies of minimalism to the patterned ground of the American deserts.

Smithson had long been fascinated with matters aerial. In 1965 he had given a lecture at Yale's architectural school about the role of the artist in shaping the environment of the city. In the audience was an engineer from the firm of Tippetts-Abbett-McCarthy-Stratton (T.A.M.S.). He was impressed enough to offer Smithson a consulting role in their contract to develop plans for the new Dallas/Fort Worth airport, a project in which Robert Morris, Carl Andre, and Sol LeWitt were likewise engaged. T.A.M.S. eventually lost the contract, but for a year and a half Smithson worked out a number of geometrically modular groundworks that would be perceived by passengers from the air. He called it "aerial art," works that would be placed between runways in the "clear zones" dictated by the FAA. His drawings of runways intersecting spirals and reflecting pools, and spiral models made of mirrors (one of which he titled an "Aerial Map") are commonly cited now as precursors to *Spiral Jetty*.

Heizer had met Walter De Maria in 1966, and brought him out to the Mojave in 1968 to help him construct the older artist's *Mile Long Drawing*, two parallel lines drawn in chalk across the desert. By this time Smithson was working on an essay for the September issue of *Artforum*, "A Sedimentation of the Mind: Earth Projects," which became the *de facto* catalog for the seminal "Earthworks" show given by these and other like-minded artists in October by the Dwan Gallery in New York. By 1969 Heizer was flying around the West looking for a new site to excavate, and settled on sixty acres on the Mormon Mesa to the east of Las Vegas, which Virginia Dwan bought for him so he could execute a commissioned work, *Double Negative*. De Maria was also out on the Mojave, this time trenching out a giant rectangle one mile by a half mile called the *Las Vegas Piece*. Although De Maria stated that the work was meant to be experienced from the ground as a walking piece through which one would come to understand the vast stillness of the desert site, it could only be seen as a drawing or sculpture from the air, and like many earthworks, its most noteworthy documentation is in aerial photographs.

In 1970 Heizer completed *Double Negative*. The sculpture, two trenches 50 feet deep by 30 feet wide along with their two debris fans of 240,000 tons of displaced rock and dirt, forms a single 1500-foot-long visual unit across a broadly eroded bay on the edge of the mesa. Invisible by foot until you're at its very edge, it has a massive and cathedral-like presence abetted by the fact that it is a cool refuge from the hot winds that blow incessantly across the top of the mesa. Seen from the air, however, it belies its origins in the mining industry. It is clearly an elegant drawing, but also a raw gesture of earthmoving. I have always found, on the other hand, that walking along the top of Smithson's salt-encrusted jetty, which he completed the same year, is a somewhat pensive experience. The work has nowhere the massiveness of *Double Negative*, and from both the air and ground seems a small gesture in the enormity of the surreal lake with its algae-stained water and a horizon so distant that water and sky merge. But it has the virtue of surprise, as an unexpected symbol that can stand for something beyond a formalist vocabulary, and as a work that appears and disappears for years at a time.

I've flown and walked over both sculptures, seeking to connect their aerial and ground views. From the air both have a structural integrity that from the ground is increasingly compromised as entropy takes its inevitable toll. The edges and debris fans of Heizer's sculpture are slowly eroding and softening. *Spiral Jetty* is alternately underwater and exposed, depending on the depth of the lake, which can vary by dozens of feet from year to year. As sediments build up around it, it is slowly being subsumed.

As for De Maria's bulldozed lines in the Mojave, I've zigzagged over their location for an hour and been unable to locate them, or perhaps simply to separate them out from what has increasingly become an overlaid grid of county and private roads as the countryside slowly fills up with a veneer of rural sprawl.

Heizer went on to start *City*, the first element of which is *Complex One*, a bunker-shaped mastaba finished in 1974, and the first of four monumental elements in a work that will be more than a mile long when finished sometime early this century. *Complex One is* large enough that it appears on USGS topographical maps with the word "artwork" on the sheet, the only such designated spot in national maps of which I'm aware. Military pilots from the Nellis AFB in Las Vegas "post" around it, using it as a landmark for making turns while practicing warfare.

Walter De Maria scouted the West from a plane to find the site in southern New Mexico for *The Lightning Field*, a collection of four hundred stainless steel poles more than twenty feet high that are situated precisely to create a suspended level plane of points one mile wide by a kilometer deep. Many of the books about earthworks, as well as the best-selling book about American art by Robert Hughes, *American Visions*, feature the piece under dark skies with lightning striking in the background as their cover illustration.

And Robert Smithson? In 1973 he was working on a relatively modest piece called *Amarillo Ramp* just outside that city in Texas. He had staked out the land for the curving and slightly ascending earthen ramp, a partial circle 140 feet in diameter that was planned to rise up to fifteen feet and be surrounded by a shallow reservoir behind a small dam built for that impoundment. Late in the afternoon he climbed into an airplane with a photographer to document the site, and shortly after takeoff the plane crashed, killing all aboard. Today the red earth ramp is slowly washing away, and despite efforts by the owner's ranchhands to keep the mesquite cleared from the piece, it's a losing battle. Stanley Marsh, who commissioned the sculpture and was immeasurably saddened by the death of the young artist, doesn't even like to talk about it.

A hallmark of what today is identified as the "heroic" era of earthworks is that the art is known mostly through documentation. Most of the large drawings and sculptures from that period were designed to be ephemeral, as with De Maria's chalk lines and Heizer's dye drawing, or have been deliberately erased and/or buried, as with the latter's depressions. Further, the more "permanent" works such as *Spiral Jetty* and *Double Negative* are located in remote places difficult for the average art tourist to visit. Pilgrimages are made by hundreds of people annually, to be sure, but most people know the work through photographs, the majority of which are aerial. More people have seen the film *Spiral Jetty*, which Smithson made upon its completion, than the actual sculpture. The film, not merely a documentary but an artwork in its own right, ends with a shot from a helicopter tracking the artist below as he runs along the top of the work to end at its center, seemingly exhausted and perplexed.

During the last several years, with the spiral out of the water, people have been leaving devotional offerings at its termination. Candles, both real and plastic flowers, little toys, and poems. The practice is akin to the attention that Nancy Holt's *Sun Tunnels* has been receiving as well. Minutes after my sighting of Lemay Island, Mike and I fly over Lucin, population zero, and the nearest named feature on maps to her sculpture.

When the first intercontinental railroad was being constructed in 1869, the Great Salt Lake was an insurmountable challenge for the finances of the time, and the route was detoured through the hills to our north. As shipping increased the added distance and slow grades caused a bottleneck in cross-country traffic. By 1902 it had become cost-effective to simply build 103 miles of new track straight across the lake from Ogden to Lucin. In the middle of the new Lucin Cutoff was a spectacular twelve-mile-long wooden trestle, an engineering feat of national fame. As with most of the great wooden railway structures from that time, it has since been replaced. The trestle, which I know of only from photographs, was abandoned in the 1950s in favor of a solid causeway, and since then the 38,000 piles and redwood planking have been ripped out and resold as salvage lumber. Given its size, the serial geometry of its structure, and its placement across the mirrored surface of the Great Salt Lake—and that it is an object we know only through documentation—I am tempted to accord it status as an earthwork.

Lucin, of course, has long since fallen into disrepair, making this exactly the sort of derelict near-past territory that Smithson and Holt liked best. It takes Mike and me a few minutes of cruising about to spot the tunnels several miles to the south, but at a quarter after noon we're circling overhead. After Smithson died in the airplane crash, Holt finished *Amarillo Ramp* as a memorial to him, but in 1976 returned here to build her own signature piece. It's one of the few earthworks from that early period that's documented more from the ground than the air, in part because it was created as a focus for people to stand inside and look upward, versus being a form meant to be apprehended as a monument. It's the reverse of the aerial, a piece that depends upon the solar and the sidereal.

If the decade of the 1960s earthworks belonged primarily, although not exclusively, to male monuments, the landed sculptures of the 1970s were often constructed by women, and tended to reflect less massive interventions to the earth, hence a shift in name for the genre during that decade to "Earth Art." Lita Albuquerque is a good example, a Los Angeles artist who started throwing dry pigments out onto the El Mirage playa north of the city in the late '70s (a site also used by De Maria and Heizer a decade earlier) and then constructed elaborate geometrical drawings that sought to manifest and thus connect celestial matters such as constellations and equinoctial conjunctions, with the ground, or heaven with earth.

Holt had bought forty acres from the Bureau of Land Management for her site, and worked with an astronomer for months to accurately situate *Sun Tunnels* and design its components. The work consists of four eighteen-foot-long concrete pipes with an interior diameter of eight feet that simply sit in an open X on the desert floor. The ends of the eighty-six-foot-long sculpture are aligned with the rising and setting sun of the winter and summer solstices. That's the solar part. Holes drilled in the tops and sides of each tunnel correspond to the positions of the constellations Draco, Perseus, Columba, and Capricorn on the solstices. The holes are drilled in four sizes, according to the magnitude of their visible stars. The piece is oriented, as is the architecture of the Anasazi and other pre-technological desert peoples around the world, toward what's happening in the sky, and to the images and narratives we construct above our heads. As Mike and I circle around the four concrete tunnels—with me holding the stick as he attempts to take snapshots of it—I realize how unassuming it is from the air. The best representations of the work are from photos taken on the ground and looking through the large tubes.

When anthropologist Robert F. Heizer was studying the rock art of the Great Basin, he declared that its petroglyphs of animals pecked into rock faces and long lines of vertical tick marks represented game and fences that would have been erected in the landscape in order to channel the game to a pre-designated killing field. Given that lithic scatter is often found in elevated sites above game trails, the idea of sympathetic magic coupled with hunting instructions makes sense. Robert Heizer is rightfully credited with bringing rigor and scientific method to the study of rock art, but it is no longer disputed that the tick marks most often are part of a solar marker indicating the solstices and equinoxes. Anthropologists used to scoff at the notion of pre-agricultural peoples keeping a calendar, reasoning that you only need to know the time of year if you were planting crops. They had perhaps forgotten that animal migrations are seasonal, and that keeping track of when various plants were going to bloom or seed was critical information for hunters and gatherers.

There's yet another major strain of interpretation for rock art, one advocated by David Whitley, a former student of Robert Heizer. Like his teacher, he bases his notion on the practices and interpretations of foreign cultures. Instead of the European cave art and interpretations of the sites such as Lascaux offered by the French and then adopted by anthropologists such as Heizer, Whitley looked at the practices of nomadic Africans, who use repetitive motion and fixation upon visual symbols to enter trance states and another world. For Whitley, the rock art sites of Western America are gateways to the underworld. I don't find all these ideas mutually exclusive. Holt's *Sun Tunnels* have, after all, been interpreted as pieces of art, as an expression of feminist consciousness raising, and are today the focus of annual solstice gatherings for neo-Druids. In terms of classifying her piece, some writers prefer to call it a site-specific sculpture rather than an earthwork, given that it does not use the earth itself as sculptural element. Actually, that's not true. It uses the Earth with a capital *E* precisely because it depends upon the rotation of the planet in order to create meaning. And concrete is assembled from earthen materials. But I get their point.

In any case, as Mike and I circle the concrete pipes from a few hundred feet above, I can see sunlight falling through the seven- to twelve-inch holes representing the stars with their apparent magnitudes, and our shadow falling across them. It's a handsome reminder of how our aerial view exists always between the ground and the vault of the sky. We are, in the airplane, not pasted onto a surface above one's head, but swimming in a three-dimensional medium. One of the greatest joys in flying is akin to that of scuba diving—the ability to slip simultaneously in all three dimensions. You can skid sideways while losing altitude yet flying forward—while traveling upside down.

This freedom to reverse the relationship of figure to ground, and looking down at a sculpture from above that's meant to be experienced while looking up, reminds me of Gary Urton's work on Incan cosmology. A Harvard anthropologist who has studied prehistoric Andean images, architectural alignments, and astroarcheology, Urton discovered that the Incas worshipped not only star-to-star constellations, but also those based on dark spaces within the Milky Way. They named the dark clouds of unlit dust for animals their shapes resembled, such as the Fox (a dark nebula between Sagittarius and Scorpius) and the Tinamou, or Partridge (the nebula we call the Coal Sack near the bottom of the Southern Cross).

Sacred spaces have been created by people around the world by orienting their architecture and monuments to the sky, which links heaven to earth, deity to human. Walter De Maria was not unaware of the tradition when he designed *Lightning Field*, a literal conductor of power from the sky to the earth. And although the last thing that Michael Heizer has in mind while building *City* is archaeoastronomy, his property and the sculpture are both oriented to the compass (although, he admits, somewhat inaccurately, based on his inexperience with the difference between magnetic and true North). The design of his mile-long sunken rectangle is based on a Mayan ball court, which was a ritualistic sport tied to their religion. And the sun does, indeed rise over *Complex One*, the first and easternmost element to be constructed within the larger design, and will set over the terminal and westernmost geometric figures at the far end.

Robert Morris, another early earthworks artist, constructed his first *Observatory* piece in 1971, an overt homage to Stonehenge, a neolithic monument the archeology of which has in no small part been conducted by aerial survey. The first aerial photograph of the site, and the first known one of any archeological site, was taken from a tethered balloon in 1906 by a lieutenant in the Sapper's Balloon Section of the Royal Engineers. It is beyond ironic that a photograph taken by a professional bombardier, whose profession was dedicated to blowing up things, would lead the British archeologists to realize the value of using aerial imagery in discovering sites to preserve.

Mike finishes his images of the *Sun Tunnels*, and we head east along the railroad line from Lucin, hugging the northern border of the UTTR until we reach the Great Salt Lake. Its size from the air is staggering. Both of us have been to the annual Burning Man festival held on the four-hundred-square-mile playa of the Black Rock Desert ninety miles north of Reno, and last year Mike flew into its makeshift landing strip. The festival is held there because its four-hundred-square-mile dry lake is the flattest and largest single contiguous piece of unimproved ground in North America, a void supreme for both contemplation and partying. Looking out over the Great Salt Lake, Mike comments: "It's more vast here than the Black Rock because of the water—the ground reflects the sky. Sort of a giant national mirror."

The Lucin cutoff splits this enormous mirror into two parts of unequal salinity, a saltier northern portion and a less salty southern one. The former part is red with algae, the latter green with plankton, and scientists now monitor the biotic activity from above with satellite images. The causeway is exactly why Smithson chose to explore the nearby eastern shoreline as a site for *Spiral Jetty*. We're not going to fly over it today, both of us having already observed it from above and visited it on the ground numerous times, yet I can't help but notice how we mentally orient ourselves to its position on the lake.

We turn south, first past the islands of the western lake, then to our right the U.S. Magnesium plant. It's a massive and rusting conglomeration of sheds, towers, and pipes that are surrounded for acres by an obviously toxic zone of sterile and discolored ground. Although it is still an active plant with a visible plume discharging from its stack, it has the appearance of an industrial ruin, exactly the sort of anti-romantic appeal that the abandoned factories of New Jersey held for the young Smithson, and that led him to pick Rozel Point on the lake as a site for his entropic spiral. We detour to take a better look, but don't get too close. This plant is the only producer of magnesium in the country, a metal used in everything from aluminum soda cans to aircraft, and since the terrorists flew into the twin towers of the World Trade Center in New York on September 11, 2001, managers of large industrial plants are wary of aircraft approaching too close.

This facility has the distinction of having been until recently the single greatest industrial source of air pollution in the United States. A byproduct of electrolytic production of magnesium is chlorine gas, which has the unfortunate effect of destroying the pulmonary systems of humans. The Germans first used the yellow-green gas in 1915 as chemical warfare, and until the British forces deployed gas masks later in the war, it was brutally effective. In 1989 the Utah plant released 110 million tons of the poison into the atmosphere annually, 90 percent of all chlorine gas in the country. In 2001 they were threatened with a billion dollars in fines by the Environmental Protection Agency. Two years later, after bitter protests, they'd brought their emission of airborne chlorine down to 2.8 million tons. The plant also produces hydrochloric acid—even in 2003 at the rate of 1.5 million tons per year. Known more commonly as stomach acid, it is capable of digesting rocks, and in fact is used for that purpose by the oil industry, which pumps it into the ground to increase extraction. Salt Lake City is downwind seventy-five miles. I'm not sure I'd be comfortable having the plant spew out more than four million tons of corrosives over my head every year.

We fly south a few more minutes, intercept Interstate 80, and turn west to begin our final leg back to Wendover. We're back in the eight-mile-wide corridor between the restricted airspaces of the UTTR. To my right is desert interrupted by an occasional industrial facility. To Mike's left is desert interrupted by an occasional industrial facility. Most of what's out here has to do with waste disposal. The plants are incinerators and storage compounds that deal with radioactive and hazardous wastes. Chemical solvents, paints, polychlorinated biphenyls (the highly carcinogenic PCBs found in old power transformers), waste from hospitals and uranium mines. The Aragonite Hazardous Waste Incinerator visible to our south burns thirteen tons of waste an hour, twenty-four hours, seven days a week year-round. Welcome to Tooele County, the nation's wasteland.

When you drive across Interstate 80, the stack of the incinerator is visible, as is that of U.S. Magnesium to the north, but you don't really see the extent to which the desert has been industrialized. It still appears like a wilderness, which the *Oxford American Dictionary* defines as "an uncultivated, uninhabited, and inhospitable region." The origin of the word in Old English derives from "wild deer," meaning a land inhabited by wild animals. Although wild deer do live in the mountains around us, it's worth noting that the closest cognate in the dictionary for wilderness is wasteland, which is defined as "an unused area of land that has become barren or overgrown." We need to update that definition as including those places that have been used to death, literally. Places to which we have laid waste, not just in the old sense of demolishing, but in terms of setting down our waste on the land.

To the south and beyond the incinerators and disposal facilities is Skull Valley, the reservation of the Goshute Tribe. While many Indian tribes around the country have made serious economic inroads by building casinos, the Mormon dominated politics of Utah have made it illegal for the Goshutes to cash in on the trend. In retaliation—and with no little sense of economic desperation—the tribe proposed to build a repository for nuclear waste, and early in 2006 the U.S. government approved licensing for them to open a dump for the most radioactive byproducts from nuclear power plants. The dump would be about forty-five miles from Salt Lake City. Go a bit farther south to the lower part of Tooele County and you encounter the U.S. Army depot where we are burning up our stores of mustard gas (the more deadly successor to chlorine gas used by the Germans in World War I) and its contemporary counterpart, sarin. More reasons not to live in Salt Lake City. When you fly over this desert what you see is a network of roads the sole purpose of which is to transport waste to remote sites for storage and disposal. You see another and parallel network built by the military in order to reach bombing target areas. From a jetliner at thirty-five thousand feet most of these lines are visible, but inscrutable, as are the bulls-eye targets inscribed by bulldozers on the ground, another kind of earthworks.

As we approach Wendover and the salt flats, we see how truly massive the potash operation is. More than a hundred miles of ditches, some up to thirty feet deep, traverse eighty-eight thousand square acres. What from the ground looks to be for all intents and purposes an infinite space available for artists to investigate and site works in is, instead, one of the most developed and tightly controlled landscapes in the world's military-industrial complex.

We're back at the Wendover airport at 1:35 P.M. and taxi up to the control tower where we tie down. The 737 that ferries in gamblers from Canada that was here earlier in the day has gone, as have the two corporate jets run by the casinos, which the executives in Las Vegas use to check up on their properties. Just as we walk away a beautifully restored vintage U.S. Navy jet, a two-seater named the Blue Angel, takes off. A trainer developed after World War II, its powerful single engine propels its two passengers steeply up into the sky. I can't help wishing I were in it. Just as the brilliant hues of the industrial facilities are toxic, yet beautiful from the air, so the military's hardware is deadly, but its ability to jump skyward is seductive as hell.

# 3. The Pit . . .

*The world beneath our wings has become a human artifact,*
*our most spontaneous and complex creation.*
                                        —William Langeweische

**The next day Mike and I drive from Wendover and across the salt flats**
to Salt Lake City, where we have an appointment with a commercial helicopter flown by Upper Limit Aviation. Our objective for the day is to fly
up into the Oquirrh Mountains twenty miles southwest of the city and
photograph the Bingham Canyon copper mine. That cavity, one of two
large excavations into the range, is two and a half miles across at the top
and four thousand feet deep, which makes it one of the two or three
largest human excavations on the planet. (The ranking depends on which
copper mine in the U.S. or Chile is currently ahead, and whose figures
you believe.) Afterwards we'll fly down to the Garfield stack at the
smelter, the tallest freestanding structure west of the Mississippi River.
Both entities are part of the Kennecott Bingham Canyon operation, the
most productive copper mine in the world.

People who write about geography, as you may have noticed, have a
thing about superlatives. It's a way of measuring the size of the world and
our effect upon it, hence our scale in relation to it. We use the tallest mountains, the deepest canyons, even the emptiest spaces, as baselines in order
to understand the size of the world and our place within it. When looking
at the largest geophysical features, or the biggest industrial, aesthetic, and

military constructions, we often require aerial technology in order to see them in their entirety, and to put them into context with the planetary surface. That applies equally to the evaporative ponds by Wendover, aerial bombing targets, and sculptures by Heizer and Smithson. Aerial images tend to reinforce the idea that it is possible to see whatever is large, and then apply it as a measuring stick to everything else, whether what is being imaged is a natural feature or an anthropomorphic one.

Protagoras, a Greek orator and teacher who lived during the fifth century BC, famously stated, "Of all things the measure is man." He was an early sophist, or philosopher who believed all things are relative to the perceiver. Contemporary cognitive scientists would have a difficult time disagreeing with his position, although most people understand that the ramifications can be disastrous if you abandon pragmatic standards for truth, beauty and justice. Despite Heizer's insistence that his work is about size and not scale, it is impossible to view earthworks without invoking scale, which is to say the size of the works in comparison to your own size. Even he admits that he makes things large in order for people to feel small next to them. And that invokes a consideration of the amount of power, or energy, it took to create such monumental sculptures. Plato, who came along the next generation and included Protagoras in his *Dialogues*, stated, "The measure of man is what he does with power." The aerial view provides a perspective on all three—man, measurement, and power—by virtue of allowing us to examine how the application of power by mankind to the earth has changed it. We do have the tendency to accept aerial views as absolutely true, privileging the elevated advantage point. After all, it takes enormous power to lift a person into the air, and that expenditure is worth more in monetary terms than standing on the ground. We tend to believe that the more effort we expend in obtaining a view, the more value it has, and therefore the more truth. It's an unfortunate fallacy.

The higher off the ground we fly in order to gaze down at the earth, the more totalizing and synoptic a view we obtain. If you're supernatural, you retain the details below and gain a perspective on the overall situation. That's what is sometimes referred to as the "God's Eye View." Being human, the higher we go the more we see of the surroundings, but the more details we lose. Aerial photography, by virtue of the fact that it can retain an immense amount of detail from high above, straddles the gap to some degree, which lends it immense veracity, especially when the views are made as vertically as possible. Part of what Mike will be doing this morning is to subvert our tendency to take aerial photographs as incontrovertible documentation of the world below by constructing aerial views that are decidedly subjective. Aerial photographs can document our effects on the planet, but they can also reveal the metrics of our anthropocentric egos, and how we attempt to exercise political and mechanical power over space and place.

If power is the measure of man, however, I would conclude that Mike and I are mildly insane. We have the right idea, to seek out by aircraft a supreme example of industrial effort; it's just that we're using a helicopter to do so, which is one of the dicier applications of power to metal that can be imagined. We're renting a Robinson 44, more commonly known as an R-44, one of the most popular and safest helicopters in the world since it was first introduced in 1997. It's a machine that's been flown solo around the world and to the North Pole, is used widely by television stations, and you can take off its doors to photograph to your heart's content. Of the more than 40,000 helicopters flying in the world, Robinson has sold in excess of 6,000 R-44s and R-22s, its two-seater model. If that sounds like an advertisement for Robinson, it's not. It's me trying to convince myself that it's not crazy to take the craft up from lake level, around 4300 feet, to above the mine and up to 10,000 feet while loaded close to its lifting capacity. We couldn't take a fourth passenger with us if we wanted to—there's not enough engine power, rotor size, or just plain air to maintain any margin of maneuverability, which, as we shall see, is critical.

Helicopters inspire me both to love and hatred. That genius of flight, Leonardo da Vinci, first drew one around 1490 after watching the spiraling upward flight patterns of birds. Who could not love a machine that allows you not just to fly across the land, but also to swoop through the air and hover over it? To fly in a helicopter is to be a fish in the air, flashing in sunlight as you spin and simultaneously change direction, altitude, speed, and both the horizontal and vertical axes of your position in the air. If an airplane attempted such a thing it would disintegrate. Perhaps that is why the following anonymous quote from the Korean War appeals so much to me: "A helicopter is an assembly of forty thousand loose pieces flying more or less in formation."

General aviation airplanes are limited by FAA regulations to staying more than 1,000 feet above urban rooftops (for all intents and purposes 1,200 or so feet above the ground in a city), and 500 feet above people, vehicles, or anything built in rural areas. Commercial planes have to stay even higher. You can fly a helicopter, by contrast, pretty much as low as you want. The primary restriction for a helicopter is that it has to be able to land in case of an engine failure on a clear surface that has about twice as much open area as the reach of its rotors. So the view you can obtain through the bubble of a helicopter and with the doors off is exhilarating. That's the love part. It's that bit about landing that's the catch. Here's the hate part. Who could fail to be distressed by the fact that if a pilot moves any one control in a helicopter, at least two more must be adjusted at the same time? Airplanes—fixed-wing craft—are by their nature stable in the air. Touch the stick to bank or roll or nose downward, and when you let go the airplane will tend to come back to an even keel. Do that in a helicopter and it keeps going in the direction you've initiated. Flying a rotary aircraft is like juggling while balancing on a beach ball. Helicopters are, in short, easier to stall.

An airplane stalls when it its wings stop generating lift, and in all but the most extreme cases, even when you're spinning literally out of control, you can correct that condition within a few hundred feet of the ground by gaining speed. You can save yourself from whatever altitude you started to drop—even without a functioning engine. Helicopter engines have at least four different ways to stall, and the model we're flying in today demands that once your engine fails, you have about 1.2 seconds to initiate the only sequence that can save your life. More about that when we're in the air.

Then there's the rate of accidents and fatalities. A 1990 NASA study found that personal helicopter flights fall out of the sky forty-four times for every one hundred thousand hours of flight, sixteen of which result in fatalities. That is, by comparison, ten times more likely than you are to kill yourself in a Cessna, or Mike's plane. A more recent study in 2004 compared general aviation accident statistics to those for civil helicopters and found that the crash rate for the latter is 30 percent higher than for airplanes. In sum, recreational and professional pilots both crash more frequently in rotary craft than fixed-wing ones. Even Leonardo came to realize that a fixed wing craft was more sensible, and by the 1500s was concentrating on glider models.

So when Mike and I pull up to meet our pilot, Sean Reid, why am I not surprised to see tucked discretely under a tarp behind a dumpster the remains of an R-22? While Mike and Sean figure out where to put people and gear, I stroll over to examine the wreckage of the smaller two-seater craft built by Robinson. I will say this for helicopter crashes. Often the people involved survive them, and that was the case here. The tail is sheared off, the rotors gone, and the bubble busted out, but the pilot and his passenger walked away from what was simply a hard landing. Perhaps "accident" is a relative term.

Helicopter pilots for commercial outfits like this tend to be young and Sean Reid is no exception. He loved flying as a kid, tried college but dropped out to get his pilot's license, and has been flying helicopters for six years. Most of the young helo guys I know were, when younger, into adrenaline-fueled sports such as extreme rollerblading and skateboarding. Sean, whose personal choice of competition was ski racing, is in his late twenties and grins easily, but is also businesslike and focused. I appreciate that. After we get Mike and his gear settled into the left seat up front, I wedge myself into the right-hand rear seat behind Sean, balancing the weight of the main fuel tank on the other side and behind us. We run through the safety procedures minus that of how to open doors if we crash—we've taken off the ones next to Mike and myself for maximum visibility. Sean gets the rotor going for a warm-up and we don headsets. Mike and I are bundled up in insulated snowmobiling suits and hats, plus I have gloves on. It will be chilly up in the mountains at ten thousand feet with an airspeed of a hundred miles per hour or so.

Once the engine temperature needle leans over into the green zone, Sean gently lifts and rotates the helicopter, always an astonishing sensation. We take off east to gain some altitude, then swing south and west to pick up Interstate 80, which we follow out of town. The four lanes of divided pavement arrow straight west and disappear into a white glare. It's about 4:30 in the afternoon and despite the hazy conditions, decent shadows are just beginning to develop around the semitrailer trucks on the freeway below us. The tower is broadcasting an audio loop: "Caution for extensive bird activity around airport," which is a hazard I've not faced before. We don't see any birds around the airport, however, or over the freeway as we head first for the tailings pile that we passed when driving into the city.

When you make a hole in the ground large enough to be seen by the naked eye from the space shuttle, you end up with a lot of dirt you have to put somewhere. One way of understanding how big a hole has been dug is to look at the tailings. The Bingham Canyon mine was an underground operation during the late nineteenth century, but in 1898 the returns were quickly diminishing. A pair of enterprising young mining engineers, Daniel Jackling and Robert Gemmell, coauthored a report outlining an audacious method by which to bring back the profitability of the property: take the mountain apart and transport the ore by rail to a nearby mill. By 1906 steam shovels were removing one hundred thousand tons of earth per month and the mountain was leveled in short order, then became a bowl, and finally the ever-deepening pit. In the meantime, a new mesa was rising up out of the pastures by the lake.

In 2006 the electric shovels were removing five hundred thousand tons per day. Since the first cut made in 1906 the miners have moved more than six billion tons of earth from which they have refined "only" seventeen million tons of copper, more of that metal than any other mine in the world. The 99.4 percent of the material that's not copper ends up as tailings, twenty square miles of which are impounded behind massive dikes more than 250 feet high. When we fly over the pile currently in use, it's so large that at first Mike mistakes it for a natural landform being carved up. We bank south over the older sections of the impoundment area, which despite their unnatural geometry are seeded and sporting native grasses. Tailings are inherently sterile, saline, and toxic. You don't want the material blowing around, and the Kennecott company has worked hard to meet or exceed EPA standards by mixing sewage (politely labeled biosolids) and wood chips into the inorganic waste, then seeding it. The motives for remediation are multiple. It's not just the avoidance of fines, but the corporate plan to develop the 93,000 acres it owns—more than half of the developable land in the valley. Suburbs already border the eastern edge of the old tailings, and the first phase of Daybreak—the largest planned community in the state—is in progress: 4,100 square acres; 13,500 homes; 100,000 trees. Once again, the only way you can grasp development at that magnitude, much less its conjunction with industry, is from the air, and as we continue to approach the flanks of the mountains toward the pit most of our flight takes place over tailings.

Just as Sean begins to pick up altitude a flock of seagulls materializes in front of us. We're in their flight path for a garbage dump visible just to the north. He immediately stops our forward motion and puts all the energy of the aircraft into gaining altitude. As Sean explains, should one hit our tail rotor, most likely we would lose not just the rotor, but its gear case, as well. It's the weight of the latter in the tail that keeps the copter in balance. Lose it and "you turn into a lawn ornament," as our pilot puts it. Lovely. Birds, however, can't climb as fast as aircraft, and we're soon clear of the flock.

We pass six thousand feet in elevation and Mike informs me that he's already taken seventy-five frames. His Linhof 4x5 Aerotechnika is no longer made, a motor-driven large-format camera used for both commercial and reconnaissance work that never had a big market to begin with, and certainly doesn't now in this age of military satellites and Google Earth. The camera is as large as, and resembles somewhat, an underwater photo housing, and Mike needs both gloveless hands to shoot with it as he braces his legs to push himself out of the aircraft door frame. It's a physical way of shooting, and afterwards he'll be both ecstatic and exhausted, much like Mary Upjohn was in Africa. If we'd lost our rotor, I wonder if he would even notice, his instincts as a pilot perhaps subordinate to those of a photographer.

We rise up through six thousand and then seven thousand feet, passing over snowy slopes and bare aspens. Gullies are shrouded in frigid shadows, and in front of us cornices hang from high on the ridges. Whenever we approach a hillside closely, the reverberation of our rotors pounds back through the cabin. Every slope under us has snow resting in deeply raked vertical lines. Erosion doesn't work this way naturally on undisturbed ground, where vegetation forces runoff into more evenly random patterns. This is chaos with a purpose behind it. Some of the ground is stripped bare and washing away because the vegetation has been killed by pollution from the mine and smelter. Some of it is overburden or tailings, so much dirt moved that it has created entire hills within the range. We're following what looks almost like a gravel glacier, but it is scoriated with a pattern of crisscrossing lines so dense it resembles the warp and weft of burlap. It is, I realize, a human raking of tailings that fill an entire valley, a pattern calculated to slow erosion by wind and rain. The leavings are, of course, sterile, and where their edge meets the steeper slopes left natural, the trees resume abruptly (Plate 2).

This barren path leads us to the first and older of the two pits, an incredibly narrow, deep, elongated hole in the earth that is, as Mike notes, distinctly vaginal in shape. The walls are orange and the terraced levels appear impossibly narrow for trucks to have driven on. Gold was sought and discovered in these mountains before copper, and this is the Barney's Canyon gold mine, once the most prodigious source of gold in Utah, but closed since 2001. I feel dizzy as I lean out of the door for a better look, but I imagine what it must have been like to drive ore trucks on those spiral ramps down into the pit. Many sections are now blocked by slumping of the rock immediately above them as the face slowly succumbs to gravity and erosion. Mike switches to a wider lens and passes the two hundred mark on his first magazine.

The pit was dug behind a shoulder of the mountain, making it look as if it is perpetually about to be squeezed shut. Because most of it is behind a ridge and invisible from anywhere but above it, it has an air of privacy, almost secretiveness. Directly above the mine the high slopes are forested and pristine in the snow. It could almost be the back bowl of a ski area, and the contrast between what appears to be an undisturbed mountainside with the severe geometry of the pit, its color and shape reminiscent of a rusted Cor-Ten steel ellipse by Richard Serra, is more than striking.

After several circuits, Sean eases the control column forward and we rise southeastward and toward the much, much larger copper pit. The terrain in between has been so bulldozed, excavated, run over, dumped on and reshaped that I can't tell if it was once a high valley or part of a ridge. Ahead of us so many stepped terraces come into view that I cannot count them, no matter how fast I try. Later, sitting with one of Mike's photos, I will distinguish more than sixty separate layers of excavation. The pit is gray in color and so outsized that it appears more like something out of a science fiction movie than a terrestrial feature. One again the roads spiraling down into the pit look too narrow for traffic, until I see a series of loaded ore trucks inching upwards, and realize how big the mine is. But that's only the first shift in scale. The second and more profound one occurs when I see a pickup truck next to one of the ore haulers, and realize that its roof reaches only to the top of one of the large machine's tires. Suddenly it no longer seems implausible that this currently may be the largest human-made hole in the planet.

The shovels at the bottom of the pit pick up 98 tons per scoop, but from our vantage point level with the rim of the pit, that's not obvious. Nor is the fact that the trucks hauling the ore upwards carry up to 360 tons of material per load. If the aerial allows us to grasp the size and pattern of the human endeavor, it also and contrarily makes it more difficult to judge the scale of it. Enough copper has been taken out of here to wire all of North America, but the pit is being expanded to meet increased demand caused by, among other factors, the electrification of China. We fly around and around, Mike shooting constantly. I try to follow the route of one of the roads, but the branching makes it impossible. According to the Kennecott website, if you stretched out all of the roads below me, they would total five hundred miles, enough to drive from here to Denver. Think of it as a stadium. The Indianapolis Speedway, one of the largest stadiums in the world, seats 257,000. The sides of the pit below us could hold nine million spectators.

The pit is a pillaged and ruined landscape feature it seems impossible for humans to have ever constructed, much less to ever restore to its former nature. Most mines can in fact be reclaimed, at least to some degree. The stripping of entire mountains in Nevada for gold and in West Virginia for coal, for example, can still be remediated somewhat by resculpting and then reseeding the remaining surfaces. They won't ever be the same landscapes, but there is a chance they will be able to reintegrate over historical time with their surroundings. Not so the pits, given their size and depth. All that will reclaim them will be the geological time scale of tectonics. You look at the pit and see for yourself why humans are now the most pervasive geomorphological force on the surface of the planet, surpassing rain as the worldwide shaper of land late in the twentieth century.

You can't look into these pits and not think of Dante's *The Divine Comedy* and its steep descent into hell. Many editions of the poem, both historical and contemporary, illustrate "The Inferno" with figures descending a circular pathway. Spiral descents into the earth, either via tunnels in the ground or whirlpools in the oceans, are archetypal challenges for humans. During previous centuries the Great Salt Lake itself had been rumored to be drained by a giant whirlpool, and some of Frémont's men expressed extreme reluctance to venture out onto it at his command. After all, what else could explain how a completely landlocked lake could accept so much runoff from the nearby mountains every year without continually overflowing? Needless to say, the evaporative water cycle was little understood by most people in the nineteenth century.

Dante, of course, offered his hero a redemptive journey, and artists have long dreamed of reclaiming industrial sites. Smithson saw a potential for ecology and industry to meet through art. He actually doodled with a felt-tip pen on a picture of the Bingham pit from a 1972 Kennecott annual report, and sketched a drawing that posited three curving tendrils on its floor, proto-spiral jetties in the toxic yellow runoff that accumulated there. It was an unsolicited idea, however, and Kennecott paid little attention. Michael Heizer, ever the iconoclast, expressed no interest in rescuing the reputations of mining companies, but appreciated free dirt, and in 1988 completed work on *Effigy Tumuli* in a state park about seventy-five miles southwest of Chicago, the former site of a strip mine. The project, which Heizer virtually disowns today, consists of five geometricized totemic figures, animals that were sacred to the Native Americans who lived in the area and constructed earthen burial mounds, or "tumuli." The catfish, frog, snake, turtle, and water strider figures were constructed of 255,000 tons of dirt moved under Heizer's direction and covered 224 acres.

*Effigy Tumuli* presents exactly the sort of problem that the aerial view can create. Many of the indigenous tumuli in America had remained undiscovered by local people at the ground level, and were only found through aerial surveys. Like many archeological features that have integrated over time into the landscape, including entire settlements such as Çatalhöyük, tumuli tend to go unnoticed when you're walking on them. They're just another hillock in the local topography. From the air, however, they stand out as artificial constructs. Heizer has always considered size to be real and scale to be imaginary, that scale is merely how we imagine the size of things relates to us, or to each other. But scale is all too real, hardwired into the way our bodies experience space. The only way Heizer could get the workers to invest in the project, and to execute his designs accurately—in short, to care—was to take each of them up in a helicopter to show them what they were working on, to give them a sense of its scale. When the Buffalo Rock State Park opened to the public, and people began to stroll on the 1.5 mile-long interpretive path along the bluffs above the Illinois River, they were unable to perceive the figures. Without being able to see them in whole, and thus bodily relate to them, no emotional bond or reverence for the site was generated. Today the sculptures, among the largest on the planet, are eroding away not unlike unsculpted tailing piles.

Sean knows that Mike is a photographer and that I'm a writer, but he doesn't know what kinds. His assumption is that, because we're spending money to rent his aircraft, we must be journalists of some sort, and as he positions the helicopter, it's clear that he thinks we want to document, or even emphasize the environmental degradation the mine has created. What Mike has in mind, however, is the construction of views that are distinctly and obviously subjective, and not simply condemnatory. He uses hand motions to deflect Sean's angle of attack so that the horizon is tilted, not level, the latter being the journalistic standard insisted upon by most news and documentary media. The skewed horizon is a way of reminding the viewer of the photograph that it does not arise from a neutral or ideal vantage point. There isn't any such thing, as every point of view is assumed by an individual and cannot be duplicated perfectly in time and space.

It's not as if you can escape the journalistic and environmental aspects of what we're doing, however, and both photography and painting have long been linked to ecological concerns in America. An oft-cited precedent for the relationship are the photographs produced by William Henry Jackson and paintings by Thomas Moran on the 1871 Hayden expedition to Yellowstone, which helped persuade Congress to declare the area the first national park the following year. Moran's subsequent paintings of the Grand Canyon had a similar effect, and the role of Ansel Adams in helping Americans perceive the magnificence of Yosemite, the Grand Tetons, and Mount McKinley is well known. But that's only half the picture, as it were, the depiction of the unspoiled wilderness. Photography, and in particular aerial photography—because it is a mechanical means supposedly not subject to the whimsy of the human hand—is accorded a visual authority even better suited to the display of the ruined landscape.

This became painfully apparent in the Dust Bowl years, which ran from 1934 through 1939, and during which the relationship of nature to photography to public was radically altered. Decades of overproduction on marginal lands, and the disruption of soil integrity by plowing up native grasses, left millions of acres vulnerable in the midwestern region where Texas, New Mexico, Colorado, Kansas, and Oklahoma conjoin. A profound multiyear drought coupled with stiff winds created thick clouds of dust that rose as high as sixteen thousand feet and blew clear to Chicago, then on to New York City and Washington, D.C. The Farm Security Administration (FSA) hired photographers and filmmakers to document the disaster. Pictures of mile-high dust clouds roiling over towns were followed by scenes of farm machinery buried up to their rooftops. The aerial photographs taken in 1936 were the start of what is now known as the Aerial Photography Field Office in the Department of Agriculture, which holds more than ten million negatives on file and is the most important visual archive of land use practices in the country.

What the storms and the photographs taught the nation was that the Dust Bowl wasn't solely a natural disaster, but one in which we had complicity. If photographs of the scenic climaxes in America helped establish parks, then the visual record of the Dust Bowl helped establish soil conservation as a national priority. Americans began to understand that to have a healthy country it wasn't enough to just hold aside areas with pretty views and forests. You had to change your relationship with the land you used in order to account for and preserve its intrinsic nature. Although photographs taken on the ground could demonstrate up close the effects of poor agricultural practices, the causative relationships between the two could be analyzed and presented in a unique way by using aerial imagery, which established large-scale relationships between contour lines and erosional patterns. The recording and analysis of land usage from the air began to be increasingly important to both policymakers and the conservationists who would influence them.

A primary example of how profound a change had occurred in the relationship among public, photography, and nature is the first book published by the Sierra Club in its "exhibit" format, a series meant to encourage passage of the Wilderness Act of 1964. David Brower, an avid mountaineer who would initiate the series, met Ansel Adams on a trail in the Sierra Nevada in 1933 when he was only nineteen years old. With the sponsorship of Adams he joined the Sierra Club shortly thereafter, worked in Yosemite Park, and then was an editor at the University of California Press for two years before being called up by the draft in 1943. His formidable climbing skills were put to work in the Army's legendary 10th Mountain Division. After the war he slowly worked his way up through the ranks of the Sierra Club until he became it its first executive director in 1952. The club had seven thousand members, most of them in California and the West. By the time he would leave the club in 1969, it would have more than seventy thousand members nationwide, a growth due in large part to the photography books Brower published. I have no doubt that Brower's climbing, and his consequent appreciation of elevated views, helped foster his appreciation of the aerial photography he would see used as a powerful rhetorical device in the cause of environmentalism.

In 1955 Bower's by now close friend Ansel Adams mounted an exhibition in Yosemite which Nancy Newhall, a noted photographic writer and historian, helped curate and design. She used extensive juxtapositions of quotes and texts with the photos in what she called "additive" captions to promote the preservation of the planet. *This is the American Earth* proved immensely popular with the park's visitors, and Brower thought the extended photo essay would make a terrific book through which the

Sierra Club could promote its goals. It was published in 1960 to almost universal acclaim.

The book's six chapters are scripted into an environmental narrative for mankind in general and the United States in particular. It begins with a two-page spread of a now-familiar Ansel Adams image, "Sierra Nevada from Lone Pine, California." The snowy Mount Williamson rises almost ten thousand feet above the valley floor, while in the dark foreground a horse grazes in a pasture. To the right stands the Mount Whitney group, the highest point in the contiguous United States. It is a little-noticed irony that the very device that made much of the growth of Los Angeles possible, the Los Angeles aqueduct, actually runs directly in front but is unseen by the viewer. The book thus starts out with a pastoral, even Edenic image, and continues with photos of birds by Eliot Porter and more totemic landscapes from national parks by Adams. The volume then juxtaposes these unpeopled images with photographs taken in the 1850s of ruins from past civilizations. One is a picture by Francis Frith of the twin beheaded "Colossi, Nubia," the other James Robertson's desolate "The Acropolis" (taken before Athens sprawled forth in the twentieth century to surround it). The meaning of both is reinforced by the texts, which remind us that civilizations fail through neglect of their environment.

By the middle of the book we have come forward in time and begun to progress upward in space, first to an elevated view of a cattle drive in the desert, then a high aerial of a log jams in the Pacific Northwest, and an oblique shot of an eroding farm in Oklahoma by Charles Rotkin. Two pages later is a stunning aerial image of contour plowing by Margaret Bourke-White, yet another professional photographer who had worked from the air during the war. The language in between the two photos makes it clear that the mechanization of agriculture divorces us from the land, and if we are not careful we are doomed to repeat the mistakes of the past. The fourth and central chapter of the book, titled "The Mathematics of Survival," is designed around five aerials taken in 1950 by William Garnett during the construction of the Lakewood tract housing development in Los Angeles. The repetitive monotony of identical houses filling the aerial frame from edge to edge created what has since become an entire genre of clichéd photos defining what we collectively label urban sprawl.

Following the Garnett photos, the book mostly returns to earthbound images and eventually ends with Ansel Adams back in the national parklands, the text a hopeful paean to mankind remembering the value of the resources around him. But the idea had been planted in the American imagination: aerial photography had a unique ability to document in fact, and forcefully convey through carefully composed views, the degrading effects that mankind was increasingly wreaking upon the country.

The book is remarkable in that it sets out the twin courses for what will become competing visions in photo books, aerial and otherwise. One branch of photography will attempt to present the world without human presence, to hold up as examples how lovely nature can be if we would just leave it alone. Much of the work by Ansel Adams deliberately excludes all signs of the human, as does that of his successor in color landscape photography, Eliot Porter. In reaction to that determined exorcism of human effect emerged what photo curator William Jenkins, working at the Eastman House in 1975, labeled the New Topographics—work by photographers that documented what was, versus that which should be. While Robert Adams was photographing suburban development along the Front Range of the Rockies outside Denver, Lewis Baltz was capturing the industrial parks of southern California.

Today there are a number of artists who treat environmental depredations as a mixture of both the beautiful and the terrible, a contemporary version of the sublime. While the earlier generation of topographers deliberately eschewed beauty as a way of countering the pristine and even sanitized pictures of Ansel Adams and Eliot Porter, the next generation accepted beauty that existed in spite of, and that was even created by, the industrial landscape. They did this, in part, by using color film versus black-and-white. Richard Misrach's photographs in his *Desert Cantos* series often include such images, the highly toxic but exquisite bomb craters on an illegal bombing range in Nevada, for example.

All of these photographers, like the earlier topographers, tend to favor large-format view cameras that use plate film to produce negatives of 4 x 5 inches or larger. That kind of equipment, usually a collapsible camera with bellows on a tripod, doesn't lend itself easily to aerial work. The compositions are studied, serious, even momentous. They bring us to a standstill for contemplation, as if we were engaged at looking at a large painting through a picture window. Mike's use of the Aerotechnika is an exotic large-format solution. It took him several years of searching online, through catalogs, and in person to find all the components for his Linhof 4x5 Aerotechnika, but it allows him to photograph with the same lush density of detail as his view camera brethren while treating the instrument as if it were an enlarged 35 mm or digital camera. As a result, there is a unique spontaneity in his work because he is able to shoot much more film, and much more quickly, than someone with a view camera. All the facts and ironic juxtapositions are there: the rapacious architecture of the mine chewing into what looks like the setting for a ski resort, for example. But his technique displays an immediacy created by what is an image obviously composed on the fly, and thus open to circumstance. This vulnerability to chance and opportunity undermines the authority of the image and allows us to question both it and what it portrays in a way that a more formal landscape portrayal would seek to prevent.

As we circle around and around the pit, Mike shoots again and again the steep ranks cut into the mountain. When I wonder out loud if we're trespassing when we fly into the airspace of the pit, he points out that this used to be a mountain; now it's a depression deeper than the mountain stood high. No matter how high off the ground we're circling here, below the rim of the pit or above it, we're flying through airspace that used to be solid ground. That's how much material has been removed. "So, who owns the space?" he shouts rhetorically. I shrug. I find myself mesmerized by the terraces, especially on the shadow side of the mine where alternating bands of snow and dirt almost read like lines of text. I've been thinking of the bottom of the mine as a small area, but when I can force myself to focus on it, I realize that it is an industrial compound with large multistory buildings and what looks to be an entire electrical substation on the floor.

The Bingham mine and smelter are complicated entities in terms of pollution and politics. According to testimony provided the Congress in 2003, the mine has poisoned seventy-two square miles of groundwater in the valley and two years earlier had released 695 million pounds of toxic waste. That figure included 21 tons of arsenic and 91 million pounds of lead, which on the EPA list made the mine the largest overall source of toxic pollution in the country. Of course, Kennecott is also the largest single taxpayer in Utah, not to mention Tooele County, and has been for a century.

We make one more pass over the older pit, this time descending below its rim and Sean pivots us around in it, then we hop over the ridge and down toward the lake. The stack pokes up above a ridge, then plays hide-and-seek as Mike communicates to Sean with hand signals how to position the aircraft so that he can juxtapose the hills with the stack for scale. It's entirely too obvious that if the pit is a vaginal opening in the earth, the stack is its male counterpart. I have no idea how the artist will address that in his photographs.

# 4. . . . and the Stack

*We no longer see the surface as concealing what is beneath it, but as explaining it . . . Flight has given us new eyes, and we are using them to discover a new order of space, new landscapes wherever we look.*

—J. B. Jackson

**We lose altitude quickly once Sean heads for the Garfield smelter and** stack. All around us everything has been laid to waste by mining. There's just no pretty way to get copper out of the ground and into our power grid. If the ground isn't eroding from direct disturbance, then the roads that are carved along, around, and into every hill have created a problem on the mountains that the mining has already stripped of trees. To our left is the wrecked mountainside, and to our right are the railroad tracks and pipelines carrying material to the smelter. It makes no sense for an ore truck the size of a two-bedroom house to haul ore four thousand feet up out of the Bingham pit, and then take it down the outside of the mountain. Instead, the 240-ton capacity trucks short-haul the ore that's blasted out of fifty-five-foot deep craters to a crusher inside the pit. The resulting rubble is dumped onto a five-mile-long conveyor belt that transports it through a tunnel and out the other side of the hollow mountain. Then grinding mills reduce the ore to a fine powder the consistency of talcum that's washed with chemicals and piped as a slurry seventeen miles away to the smelter, where it's cooked at 2,500°F. And that's the process that causes much of the pollution from the Kennecott operation. Copper comes attached to sulfur, and when the copper is cooked, it produces sulfuric acid, three thousand tons of it each day.

The original smelter was built by another company in 1905 and had smokestacks only about three hundred feet tall. Kennecott bought the smelter in 1959 and started on a series of modernizations. Nevertheless the smoke, which by 1970 was dropping forty-eight thousand tons of sulfur dioxide over the surroundings per hour, was so toxic that it killed all the vegetation for miles around and high up into the mountains behind the refinery: The bare naked mountains over which we're now flying. Sean pilots us north from the pits and around a small ridge with just a trace of snow on its upper lip, cutting a diagonal from pit to stack that shortcuts the slurry pipes.

The Clean Air Act of 1970 made the concentration of emissions an untenable situation, but it was not without protest that Kennecott bowed to the inevitable and spent $300 million to further modernize the smelter and erect the tallest freestanding structure west of the Mississippi, the Garfield Stack. That's the smokestack over the mountains, around which we are still flying. The top of the stack itself is just now peeking over the ridge.

The new stack was still putting out enormous amounts of sulfur dioxide, but dispersing it over a much larger area in a diluted form. In 1995 Kennecott spent another $880 million on the smelter, making it the cleanest in the world, at least according to the company. The refinery now recaptures all but a few pounds hourly of the sulfur as acid, which it produces to the order of a million tons per year. We are, apparently, in no danger of suffocating ourselves in a plume of poisonous gas as we fly around the stack. Kennecott has the pleasure of not only claiming to have cleaned up its act, as it were, but is also now marketing commercially viable sulfuric acid.

We bank around the ridge, Mike shooting away, and I notice several things. First, the stack is taller than even some of the surrounding hills. For decades it's been the most notable structure in the desert when you fly into Salt Lake City, much less when you drive by it on I-80. Yet it's doing nothing but becoming more impressive as we fly toward it. No matter that we routinely fly thirty or forty thousand feet above the ground in commercial jetliners; there's nothing like a tall edifice to command awe, respect, even worship. Humans have long attempted to touch the sky by erecting ever-taller structures. No one has any idea if the Tower of Babel existed, much less how high it might have reached, but if it was inspired by or was even an actual Mesopotamian ziggurat, it's worth noting that the best Nebuchadnezzar could do some two thousand years later was a stepped pyramid less than four hundred feet high. The Great Pyramid in Giza, finished around 2560 BC was originally 480 feet high, but when the Pharos Lighthouse of Alexandria was built in the third century BC, again topping out around 490 feet, it remained the tallest structure in the world for centuries. Its three-tier design of a square section on the bottom, an octagonal middle, and a circular top was the basis for early minarets, and when visited in 1183, the lighthouse actually held a mosque on top. During the early 1300s, earthquakes demolished most of the structure, and the great cathedrals of Europe would next claim the honor of being the world's tallest structures. Sky and ground, heaven and earth: the same story.

The Garfield stack has always evoked those comparisons for me. The gently tapered circular tower is 1,215 feet high, which means it is just 35 feet shy of the New York Empire State Building, and is taller than all but eight other buildings in the world. Its upper courses are so far from the ground that they had to be poured from vats carried aloft by helicopters— the builders just couldn't pump concrete that high. Beyond its height is the fact that it stands in the desert with no other buildings in competition, unlike the skyscrapers of New York, Chicago, Taipei, and other cities. This stark industrial gesture rises from the shores of an ancient lake, while behind and to its left are the stacked terraces of the upper mine pit. It's an archetypical male-female juxtaposition of forms, and in my imagination evokes pre-Biblical stairways to heaven. It doesn't hurt that it's within sight of the Mormon Tabernacle, which I've always considered an architectural symbol of similarly anachronistic religious impulses.

It's true that the stack is considerably shorter than several dozen broadcast transmitting towers around the world, structures supported by guy wires of which the tallest is in Blanchard, North Dakota. It tops out at 2,063 feet. The Warszawa radio tower that the Poles constructed in 1974 was even taller, reaching up another fifty feet, which would have made it still the tallest manmade structure in the world, but it collapsed in 1991. Radio and television towers, however, don't really count in this contest. They're spindly erector set affairs, more engineering than architecture, while the concrete stack is an actual building with an enclosed interior structure that is 120 feet in diameter at the bottom. Neither is the stack an empty tube, but contains various vertical elements, among them the reinforced duct that carries the smoke upward, and an elevator that climbs up and down a geared track. The trip takes ten minutes each way.

Sean swings us to the east and then north of the refinery and its stack, then at Mike's direction starts spiraling around it and slowly descending. We start out above the top of the stack and drift downward to where we're even with the Bonneville Bench, the most prominent of all the ancient shorelines left behind by the Pleistocene lake and about nine hundred feet above current lake level. To the south stand the snow-covered Oquirrh Mountains, but the rest of the basin is still obscured by haze in the late afternoon sun. Everywhere within sight, and the ground is flat, earth has been displaced, planed flat, segmented. There's not a tree in view, although Kennecott boasts it has planted 150,000 of them in recent years.

Mike keeps making gestures with his right hand to get us closer and lower, all the while clicking away with his left. He's pushed himself as far out of the helicopter as he can, and if the seat belt gave way he'd almost instantly become part of the industrial landscape below. Sean gets us down to five hundred feet off the ground and as close to the stack as is feasible, enough so that we can see not only the individual courses of concrete, but also the wavy vertical lines in them that show how they dried (Plates 3 and 4). I could throw a tennis ball out the doorway and hit the stack. Instead I try to do something productive, like take notes, but mostly there's just too much to look at, and I know I'll mostly be relying on Mike's contact sheets as a visual record.

The west side of the stack is in the light, the east in shadow, and I know that the black-and-white prints will separate the two more so than in reality, our color perception somewhat overriding boundary contrasts between light and dark. The smelter's pipes, vats, and ground are much cleaner than the U.S. Magnesium plant we'd flown by yesterday. Across the freeway and lake, Antelope Island is dimly visible.

Mike's jamming himself out the door reminds me of a wonderful photograph made by Margaret Bourke-White in 1951. She's hanging out of a
helicopter in a harness attached to a cable, working an assignment for *Life*
magazine. By that time she was the publication's official aerial photographer, and that same year she became the first woman to fly in a B-47 jet,
part of the Strategic Air Command out of Wichita, Kansas. She was at the
height of her powers, one of the most famous photographers working in
the world, and her specialty had always been industrial structures. She
would have loved this flight.

Bourke-White was born in 1904 to an engineer working in New York City
who kept a set of drafting tools at home in order to work there in the
evening. Both of Bourke-White's parents were keenly interested in natural history, and so was their daughter. Her first major in college was her-
petology at the University of Michigan, the beginning of a peripatetic
course of studies that took her through paleontology at Purdue and zool-
ogy at Cornell. But in 1921 she was at Columbia studying photography
with Clarence White, a leading "pictorialist" of the time. Bourke-White
opened her first studio in Cleveland in 1927, deep in America's industrial
heart, and at first made a slim living with picturesquely softened architectural photos. But that was the year Lindbergh made his flight across the
Atlantic, and Le Corbusier's modernist manifesto *Towards a New Architecture* appeared in English in America, a tract promoting the building as
machine and holding up the American skyscraper as its apotheosis. The
Museum of Modern Art would open its doors in 1929, and the Empire
State Building would be completed in 1931. The painter Charles Sheeler
had already visited Europe twice and by the early '20s he was making a
living as an architectural photographer of skyscrapers, and taking straight
photos of industrial structures. The same year that Bourke-White moved
to Cleveland, Sheeler was hired by Henry Ford to photograph his Dearborn automotive plant.

Bourke-White found herself attracted to the ubiquitous steel girders
and smokestacks of industrial Ohio, and began taking their portraits
within months of founding her studio. In 1928 she talked her way into the

Otis Steel Company plant, and after working there virtually every night for five months became the first person to successfully photograph the various stages of making of steel. Her pictures were widely reproduced and their outright theatricality—borne out of a straight technique versus a pictorialist one—caught the attention of Henry R. Luce, the publisher of *Time* magazine. He invited her to New York for a meeting, whereupon he hired her half-time as the first staff photographer for his new magazine startup, *Fortune*. She was all of twenty-five.

After Sheeler's photos of the Dearborn plant were published, Luce hired him to paint a series of six works with industrial power as their theme. The first issue of the magazine appeared in 1929 with three photo spreads by Bourke-White, and that winter the editors asked her to photograph the construction of the new Chrysler Building, which ever since its completion has been admired as one of the classic skyscrapers in the world. Among her vantage points was one on the sixty-first floor where the famous steel gargoyles projecting outward gave her an airy perch from which to photograph eight hundred feet over the city. She liked it so much that she put her studio there when the building was finished.

In 1930 she was sent to Germany to photograph its industrial plants, a trip she managed to extend to Russia, where she did the same, the first photographer allowed to do so. She returned to Russia twice and published a book about its industrialization and people, *Eyes on Russia*, in 1931. Along the way she helped invent what we now call the photo-essay, still the dominant narrative form of photography in print media. In 1934 she was assigned by her editors to document the effects of the American drought. She had a deadline of only five days, and once in Omaha realized that the territory of the Dust Bowl was so large the only way she could cover the story coherently as photo-essay was by airplane, the same method the FSA was using to study it. She rented a small plane and a pilot, flew the Midwest from Texas to the Dakotas, and despite a crash landing on the last day brought back images that showed the encroaching dust swallowing the heartland of the country. The next year, supplementing her Depression era wages from Luce, she started freelance work for various airline companies, Eastern Airlines most prominently among them. She worked over Manhattan and photographed a DC-4 below her as it flew past the Empire State Building, in one composition capturing simultaneously the power of industry and aviation and the vertical architecture that symbolized corporate America.

Luce started publishing *Life* magazine in 1936, a periodical devoted to the photographic essay, and gave Bourke-White the cover story, a spread on the construction of the Fort Peck Dam in Montana. She photographed from the air in low oblique shots taken in slanting daylight. The dam, ever a demonstration of industrial might over nature, was thus put into the larger contexts of both nature and society.

Bourke-White was by far the most acclaimed industrial and aerial photographer of early- to mid-twentieth-century America, and in 1942 was accredited as the first female war correspondent and official U.S. Air Force photographer for *Life*. It probably didn't hurt that one of her best friends was Edward Steichen, who had not only been the commander of the photographic division for the American military in World War I, but during this new conflict would become the director of the U.S. Naval Photographic Institute. It took her a year of pleading with the brass, and working aboard a troop ship that was torpedoed out from underneath her, but she finally got General Jimmy Doolittle to grant her permission in early 1943 to fly with the 97th Bombing Group's B-17s on a bombing raid over Tunis. She was first flown to an oasis in the Sahara by a young pilot named Paul Tibbets, who less than three years later would take a B-24 named after his mother over Hiroshima to drop the first atomic bomb. Bourke-White flew out of the oasis and over the bombing of an airfield from fifteen thousand feet, taking pictures of both the ground and the flak bursting in midair around her.

It wasn't all aerial work for Bourke-White. As always when creating a photo essay, she also worked on the ground, understanding that the aerial, while it revealed the larger context for a story, tended to diminish the personal. By keeping her feet on the ground she exposed herself to the same physical and emotional hazards that the civilians and soldiers experienced. She was shelled several times, documented the dying and wounded in hospitals, and in 1945 accompanied the troops liberating Buchenwald. Toward the very end of the war the Air Force commandeered her services to take aerial photos of Germany's industrial cities so they could gauge the results of heavy bombardment over cities such as Nuremberg. Which is, eventually, how she got to fly in 1951 with the Strategic Air Command out of Omaha and ended up suspended under a helicopter.

Bourke-White was diagnosed with Parkinson's disease in 1952, but that year was still flying over both coasts on assignment, often in helicopters, even having to be rescued from a crash into the Chesapeake Bay. Later in the decade she was working in the air over farms in South Korea and Colorado, and was assigned by *Life* to be the first photographer on the moon, should the opportunity arise. By the time she died in 1971, Bourke-White had become not only a photographer of the famous, nor just a famous photographer, but also a celebrity in her own right. And that gave her an astonishing pulpit from which to promote aerial photography as both an important journalistic tool and as an aesthetic genre.

Bourke-White also was a major figure in effecting an important change of vantage point toward the machine. Prior to and during the Industrial Revolution, machines were literally and figuratively looked up to. Whether it was Jacob von Ruisdael painting waterworks in the seventeenth-century Dutch countryside, or Sheeler photographing the Ford plant in 1927, the vantage point was usually below the mechanism, part of a shift from worship of things unseen, the hand of God, to the increasingly visible works created by the hand of man. That held true for Bourke-White when she was photographing the Fort Peck Dam and the massive Soviet industrial installations. The machines loom over workers, the viewer, and humanity. The most common early depictions of the Eiffel Tower, the first use of steel to erect a public vantage point over a city, were from street level and a celebration of how it connected the pedestrian to the sky. But almost immediately people were taking pictures from the top of the tower and the skyscrapers that soon followed. Bourke-White, by photographing from one airplane down onto another, both of them over the Empire State Building, turned our gaze upside down and lowered our expectations. Now the tall buildings were part of an urban structure, a human design. Instead of connecting Earth and Heaven they were more about fusing the ground with the heavens as a consolidated medium of transportation and industry.

* * *

By the time I've scribbled down the notes to remind myself to bring Bourke-White and precisionism into the discussion, Mike has asked Sean to cross the freeway and fly us over the wetlands bordering the Great Salt Lake. Because the lake is so large yet so shallow, and in a desert—which means, perversely, that the environment is more shaped by intense precipitation events and by flooding than any other natural factors—the salinity and chemical composition of its waters vary considerably from place to place. Whereas the waters up north by *Spiral Jetty* tend to run red with algae, here they are blue and green. People are out walking along the shoreline. There's no fishing in the lake; it's far too salty for anything but brine shrimp, but any large body of water will attract humans along its edge.

We're close to Saltair, and decide to circle it. This is the third incarnation of the Moorish-styled resort, the first being built by the Mormons in 1893 as a family attraction. It rested on more than two thousand pilings to keep it dry above the fluctuating lake levels, and it floated mirage-like above the briny waters like an American version of the Taj Mahal. It was a premiere attraction for two generations of locals and tourists alike who swam outside during the day and danced inside during the night, before it burned down in 1925. It was rebuilt, but there was another fire, then the lake receded during a drought, and after World War II there were other entertainments closer in town. The crowds dwindled away, and yet another fire in 1970 destroyed the second building. The current version using a salvaged Air Force hangar was raised in 1981, only to find itself flooded as the lake rose to record heights during that decade. The owners dried it out, and after a period of virtual abandonment it now serves as a rental hall for rock concerts and other events. From the ground level as one approaches its front entrance, you're flanked by palm trees and can feel at least an echo of the building's Moorish origins. From the air it's just another flat-topped industrial building with a large parking lot.

That's always been a traditional tension between ground- and air-based views of buildings. We have, for the most part, built structures to present a face, a facade, to a person approaching on foot. The aerial view of most buildings is at best a few simple geometries covered in tile or shingles. At worst, roofs are barren tarpaper expanses interrupted only by the sheet metal of air conditioning units. Google Earth is changing all that. As more and more of us look down on our homes and businesses, it occurs to us that this is now a public face, as well. People are planting gardens on their roofs in record numbers, not just in response to global warming and energy conservation, but in order to position themselves more aesthetically to the world view. Saltair and other entertainment venues would do well to remember the signage that farmers used to put on the roofs of their barns, letters advertising everything from tobacco products to produce that were so large they were visible from jetliners flying thirty-five thousand feet above them.

The marshlands that we're flying over appear healthy from this altitude, belying the toxic plumes that run underground from the nearby tailings pond. A flock of geese are drifting slowly about through the reeds. There's an obvious photo that Mike avoids taking, the one with the wetlands in the foreground, the stack and cooling towers of the smelter in the midground, then the barren foothills, everything nicely backed up by the ironically pristine snowy mountains. Like many photographers with a training in art, he avoids the easy eco-porn shots that a magazine editor might drool over. The presentation of the human-altered landscape through aerial photographs has many branches, some of which are overtly rhetorical, others of which deliberately eschew all commentary in order to allow viewers to make up their own minds about what they're seeing. It's worth remembering that the aerial photos made by William Garnett over the Los Angeles Lakewood development in 1950, then used by Nancy Newhall and Ansel Adams in *This is the American Earth*, were actually commissioned by the developer to show with pride how he was using a manufacturing economy of scale in order to provide houses that the parents of the postwar baby-boom generation could afford. As with all images, intention is everything. The intent of Garnett framing a housing development in order to emphasize the efficient replication of a basic form was subverted to the intent of an editor, Nancy Newhall, to point out that the repetition has a horrifically numbing effect on our relationship to land.

The Center for Land Use Interpretation, which is a primary nexus for the meeting of aerial photography, an aesthetic practice, with geography, a scientific practice, hosted an exhibition in 1999 of the oblique aerials of open pit mines in Nevada taken by Bill DuBois. Most of them were of gold mines, sixteen "monuments of displacement," that at first glance looked more like excavated Aztec or Mayan temple cities than the industrial ruination of entire mountain ranges. DuBois has been taking these pictures since 1975 from his Cessna, snapping away with a medium-format Hasselblad while piloting his plane more than three hundred thousand miles around the Great Basin. He worked as a mine inspector for the State of Nevada for years, then switched to managing a mine in Aurora. CLUI presented the exhibition without commentary, and hosted a lecture by DuBois where he talked about how the gold was stripped from millions of tons of dirt by washing the heap leach piles with a mixture of cyanide. The point was to provide information so that the audiences for the show and lecture could make up their own minds about the relative value of gold used to make watches and electronics, versus desert viewsheds free of glaringly obvious anthropic intervention.

The intent of David Hansen, another aerial photographer, is to document mining sites with the stated purpose of exposing their toxicity, but at times you can't distinguish between the aesthetic effect of his photographs with those made by DuBois, who is celebrating the industrial power of some of the same sites. Hansen is, at the beginning of the twenty-first century, the photographer who perhaps best exemplifies how the topographical aesthetic and aerial technology are well-suited to each other and environmental causes. His book *Waste Land: Meditations on a Ravaged Landscape* was published by the Aperture Foundation, which was founded in 1952 by six people prominent in American photography, among them Ansel Adams, Beaumont Newhall, and Nancy Newhall. The mission of the foundation is two-sided: to promote photography and the exchange of ideas among people looking at photographs. Hansen is an excellent fit. A native of Montana and witness to the long-term depredations of mining in his home state. The book, published in 1997, contains four series, the first and earliest from 1983 to 1985, sixty-six color photos of Colstrip, Montana, one of the largest coal strip mines and power plants in the country. The series begins with establishing shots of the complex made from ground level, but ends with aerial views of the larger geography, which places the mining, power generation and transmission, waste byproducts, and company housing into context with one another. The photographs are by turns handsome documents of industry and shocking indictments of pollution, their colors ranging, respectively, from the delicate grays of late winter afternoons to the acid greens and yellows of waste ponds. The aerials of deforestation and subsequent excavations display a penchant for the abstract, even as they plot out the scope of an environmental disaster.

But it is the central series of the book that has become a touchstone for students of cultural geography, which unlike its counterpart, physical geography, focuses on what humans do to the land. In 1982 the EPA inventoried approximately 400,000 waste sites in the United States. Hansen selected 67 of the most toxic 782 of the sites, and in 1985–86 documented them from the air. Each site takes up a double-page spread and consists of, from left to right: a segment of the relevant USGS topographical map with the site boundaries indicated; the official EPA site description; and, finally, the aerial photo, an oblique that tends to put the site into literal and meaningful local context. Some of the sites are infamous; Love Canal in Niagara Falls, New York, is an example, which from the air is an anomalous and relatively innocuous looking strip of cleared land cut through the middle of a suburb. Another site is not far from where Mike and I are working today, the North Area of the Tooele Army Depot where aging munitions are stored and "demilitarized." But others are totally obscure to the general public, like a small waste area outside Baton Rouge and next to the Mississippi River, or an abandoned asbestos mine tucked into the hills of Fresno County, California. These are sites you would never stumble across or be allowed to see were it not for exposure through the aerial view.

Hansen writes as follows in the his introduction to the book about his use of the aerial:

As my work evolved, the aerial view increasingly seemed to be the most appropriate form of representation for the late twentieth-century landscape: an abstracted and distanced technological view of the earth, mirroring the military's applications of aerial photography for surveillance and targeting. The aerial view realizes the Cartesian rationalization and abstraction of space that has preoccupied Western culture and visual art for the past 300 years. It delivers with military efficiency a contemporary version of the omniscient gaze of the Panopticon, the nineteenth century's ultimate tool for surveillance and control. The aerial perspective also allows for the framing of relationships between objects that may seem unrelated on the ground, and it permits access to sites with security restrictions. What otherwise cannot be pictured becomes available to the camera.

The aerial vantage point also, and not at all coincidentally, keeps him out of harm's way. It's not that exposure to any one site would be fatal, but to work for a number of years at ground level on multiple toxic sites, each with its own unique stew of poisons—on top of exposing oneself by necessity to the multiple chemicals inherent in the business of photography—seems foolhardy. The book is carefully edited to give the impression that "things are what they are:" in the effort to construct a human society, we deconstruct an ecosystem. The jeremiads are left to the accompanying texts by notable environmental writers Wendell Berry, Terry Tempest Williams, and others. The juxtaposition of the maps with our granting authoritative status to the aerial view, coupled with the passion of the writers, makes for an effective document.

As we turn east toward Salt Lake City and parallel the long embankment of the tailings, I think about Hansen's book and the photo that Mike didn't take, the one that would have juxtaposed so neatly the wetlands with the stack. It's all about intervening ground, and what in a few years will perch between our position just north of I-80 and the Bingham Pit, that master-planned development of Daybreak. It will start with 4,100 acres; 17,500 homes; 100,000 trees. Kennecott owns altogether 93,000 acres here, more than half the developable land left in the valley, from the top of the mountain to the other side of I-80. The aerial images will be irresistible. The last photos Mike takes are of the salt flats west of the airport, a strange no-man's land that goes wet and dry as the lake rises and falls, and that is covered by inscrutable straight lines representing everything from industrial to recreational use.

On the way back to the airport I revert to my morbid fascination with how helicopters manage to stay in the air, and ask Sean about the glide ratio of a helicopter. Let's say you're in a 747 flying at thirty-nine thousand feet and all four engines quit. That's never happened, but if it did, the aircraft could glide for at least 110 miles before touching down, a ratio of around 15:1. The best sailplanes can get around 60:1, and even Mike's little eggshell manages a very respectable 18:1. A helicopter? 4:1. If a helicopter engine fails, the only option is to auto-rotate the rotors, letting the airflow as you fall keep the blades turning to slow your descent.

"Want to try it?" asks Sean. Fool that I am, I say "Sure." He switches off the engine and we begin to plummet. Well, almost. It's a very steep glide that would produce a heck of a hard landing, but probably not be fatal. I think I'm reassured. He reengages the engine, but just to make sure I get it, does it one more time while on approach to the airport. When we land, I get out shaking my head, looking over at the remains of the destroyed R-22. We've been in the air for an hour and a half and Mike has taken more than four hundred photos. Given the rental price of the helo, that's a little more than $1 per picture, in Mike's mind a cost-effective way to work. Once he edits and sequences the images, he'll settle for a group of twenty-one photos handbound in his studio into a book titled *Bingham Mine/Garfield Stack 04.21.06*. The book, one of only ten copies made, will be printed on an archival matte paper, and you'll take it out of its custom-made box and then while wearing white cotton gloves open wide the pages to their double-page spreads of 36 x 44 inches. The book is designed to open to about as wide as a normal person can stretch his arms. You'll thus be looking not at a printed reproduction, but actual photographs that have been bound into a codex form: that assemblage of pages with a spine that encourages us to go from one double-page image to the next. It will be a linear narrative of your flight, an exploration of a world from above even fewer people would ever witness otherwise. If you were to put such an object on a shelf, it would sit comfortably with the expedition reports illustrated by Timothy O'Sullivan, Mary Meader, and the *Apollo* astronauts walking the lunar surface with Hasselblads strapped to their chests.

# 5. The Horizontal City

*Flight's greatest gift is to let us look around, and when we do, we discover that the world is larger than we have been told and that our wings have helped make it so.*

—William Langeweische

**One week after flying over the Bingham mine, Mike and I meet Denis** Cosgrove at the CLUI building in Los Angeles for a late afternoon helicopter tour out of Long Beach. Denis is a slight man with thinning hair and a winsome air, a preeminent British geographer now teaching at the University of California Los Angeles who specializes in the relationships among maps, aerial views, and culture. We've all worked together before in different combinations, and have wanted to fly as a trio for at least a year. Denis is currently looking at aerial representations of the city during the twentieth century. It's his conviction that because L.A. is so large and geographically dispersed, as is his home city of London, it can only be fully understood as a city from the air.

Denis is also a helicopter virgin, and we're looking forward to witnessing his reactions to the flight with his door off. As with Mike, Denis's father was a bombardier in World War II. Last summer I walked around a working-class neighborhood in Islington, the borough of central London where he maintains a flat. It's a section of the city that's changed considerably since its housing boom in the nineteenth century, when it became the densest residential district in the city. He pointed out the older three-story houses that had been divided into various combinations of townhouses and flats during the 1880s, and then the obviously much newer construction erected on lots where German bombs and rockets flattened 3,200 residences during the war. Just as his neighborhood has changed since mid-century, so he believes has Lakewood, the tract development photographed by William Garnett. We'll check it out this afternoon from an altitude close to Garnett's when he portrayed it.

We leave CLUI for the airport about quarter of three, the traffic on the freeways headed south from Culver City to Long Beach about as bad this Friday afternoon as you'd expect, so we cut over on surface streets through the Baldwin Hills. The low-slung ridges run just north and east of the Los Angeles International Airport, and they're dotted with oil rigs and derricks dating as far back as the discovery of the Inglewood Oil Field here in 1924. I note that many more drills are active and pumps working than several years ago when I was first writing about the field. The price of oil has gone up since then.

Why L.A. is so large and spread out has everything to do with its specific geography and underlying geology. First, let me define the foolishness I'm committing when I write "L.A." and "city" in the same sentence. The city of Los Angeles covers 498 square miles and holds roughly four million people, but the boundaries of this largest single civic entity inside Los Angeles County make it a complex piece of a puzzle defined by that larger body. Said county covers 4,752 square miles with ten million people living cheek by jowl. In turn, the combined urban statistical area includes five counties and eighteen million people. All of it is set inside a megalopolis that stretches almost uninterrupted from Tijuana to Santa Barbara. This contiguously built environment is larger than the country of Ireland.

So, for purposes of comparison, not to mention sanity, I usually keep my facts and figures about the place limited to the county, because when you're in the air over what locals call L.A., that's mostly what you're looking at. Its population is larger than forty-two of the fifty states, and includes eighty-eight separate cities, most incorporated and some not, all referred to as Los Angeles, much to the annoyance of some residents in the San Fernando Valley who are forever attempting succession from L.A. The county has four rivers, four mountain ranges, and ranges in elevation from 9 feet below sea level to 10,080 feet above it. On a clear winter's day you can stand amongst palm trees above the Pacific Ocean in Santa Monica and see the snow on Mount San Antonio some sixty miles away, all in the same county. The metropolitan area covers an enormous and topographically complex area, and L.A. is the only major city in the world divided by a mountain range—the Santa Monicas and their tail, the Hollywood Hills, separating the L.A. Basin from the San Fernando Valley. It is almost impossible to function as a resident in it without a vehicle.

Most of the city of Los Angeles is built within the L.A. Basin, a depression thirty thousand feet deep filled with sandy alluvium from those mountains, the steepest in North America. And those sands are filled with oil that's been pumped steadily since 1892. Drive around the city and you will see not just the naked derricks of the Baldwin Hills, but rigs disguised as office and apartment buildings in Century City and behind the Beverly Center mall on the edge of the Beverly Hills. More than a billion-and-a-half barrels of oil still reside under L.A., and although it's not all recoverable, it's still one of the larger reserves of petroleum on the continent. The original oilfields were scattered all around the county, part of the reason why the urban grid of L.A. grew so large. The city expanded to encompass more drilling sites early in the last century, which made L.A. once and still the most car-intensive city per capita in the world. When oil first gushed up from a well downtown, the population of the county was about 100,000 people who walked and rode in horse-drawn conveyances over seventy-eight miles of streets. Just over twenty years later L.A. had 55,000 registered vehicles, already making it the most auto-invaded city in America. In 1920 L.A. hosted 936,455 people who owned more than 600,000 autos.

L.A. County now has 21,253 miles of streets and roads, of which 527 are freeways. Its longest street, Sepulveda Boulevard, runs 76 miles from Long Beach to the north end of San Fernando Valley and takes at least two hours to drive from end to end. More than seven-and-a-half million vehicles are registered in the county, and every day Los Angelenos commute more than ninety million vehicle miles in them. L.A. is defined by the car, and you inevitably will experience the city through a windshield, which organizes it into a linear narrative. But if you want to grasp the *gestalt* of this place, the scope and nature of its geography and what the automotive experience of it means, you need to obtain a synoptic view. You have to see it from above.

Right at four o'clock we pull into the parking lot of Los Angeles Helicopters, one of the largest firms flying in the region. Among the services they offer is support to various law enforcement agencies, and a staff photographer who does assignments for law firms and real estate companies, construction and oil firms, private investigators—and people who want to propose to their dearly intendeds while cruising above the beaches of Santa Monica at sunset. Today, however, we're asking them to provide an aerial platform for Mike as he attempts to expose color film late in the day under a watery, overcast sky and in hazy conditions. It's about the least romantic atmosphere you can imagine for a photographer, but a specific condition essential to Los Angeles in several ways.

That the Los Angeles Basin is such a large and flat receptacle for marine air, and that it is surrounded by the forever rising and eroding San Gabriel, San Bernardino, and San Gorgonio mountains, are the conditions that produce the uniform, soft, and malleable natural light required by the film industry before the use of artificial light and indoor studios became feasible. Los Angeles is a mostly sunny, semiarid, warm place next to the cooler Pacific Ocean. This produces daily onshore breezes that push inland a marine layer that thickens in early summer to produce that foggy weather known as "June gloom." In addition to the humidity, the granite in the surrounding mountains, as it weathers, decomposes into a fine dust, the particles of which tend to be about a micron in diameter. That one-millionth of a meter is close enough to the wavelength of visible light that it helps scatter and diffuse it into a palpable luminosity. Great for the filmmakers, less so for most photographers, but Mike often gets around that by shooting into the sun, which in a clearer atmosphere he would be unable to do. The resultant aerial photographs of Los Angeles are dissolved by what painters call "halation," where light leaks everywhere onto everything. It's the same technique, albeit done with photographic emulsion instead of with paint, that J. M. W. Turner used to dissolve his landscapes in a bath of light as he painted facing the sun.

* * *

Andre Hutchings is our pilot today, an Aussie from North Melbourne. He's been flying helicopters for fifteen years, and was formerly with the Baldwin Park police department, part of the reason why the company has strong connections to local law enforcement agencies. Just as Los Angeles has long been a city defined, even dominated by, automobiles, so it has been one connected to the airplane. To give just a few examples: The first American International Air Meet was held nearby in 1910, only two years after the Wright brothers flew at Kitty Hawk, and by 1912 airplanes were being built in the city by the men who would go on to build the foundations of the aerospace industry. Glenn Curtiss, who organized the air race, built the first plane to fly from a ship, then designed a "flying boat," what we now call a seaplane. Working with the U.S. Navy, in 1919 he built the first seaplane to cross the Atlantic. Lawrence Bell was also working in L.A. at the time and went on to found the Bell Aircraft Corporation, which built fighters in World War II, the first jet-powered aircraft for the Air Force, the first plane to break the sound barrier, and, starting in 1941, the helicopters that would make the machines a ubiquitous presence over Los Angeles.

By the 1920s the *Los Angeles Times* found it necessary to publish bird's-eye views in order to explain the rapidly expanding city to its residents. That same decade the Los Angeles Police Department (LAPD) began to realize that L.A. was too big for its officers to patrol effectively on foot. What the agency needed was a way for a central command to dispatch officers quickly anywhere in the city. The answer was radio patrol cars, among the first in the nation. But even that wasn't enough, and in 1931 they hired their first fixed-wing pilot. That was great for observation, but not tactical involvement, and in 1957 the department became the first in America to field a police helicopter over an American city. The most active local television station, KTLA, kept pace by sending up the first TV news helicopter the next year.

The LAPD estimates that an officer in a helicopter can surveil more than fifteen times the personnel in a car, and they now fly seventeen of the aircraft. The fire department flies five of its own, and the Department of Water and Power (DWP) runs four over the Los Angeles Aqueduct and other facilities. Since the terrorist flights of September 11, 2001 DWP flights have increased fourfold. The LAPD helicopters put in more than seventeen thousand hours a year, at night turning crime scenes into stark stage sets with thirty million candle spotlights. In 2005 local law enforcement agencies started deploying Unmanned Aerial Vehicles (UAVs) over the Oscar Awards. The tiny craft can maintain station over a scene for hours on end to feed real-time, high-definition, securely encrypted aerial images to mobile command centers from thousands of feet higher, and with much greater standoff distances, than the helicopters. They are for all intents and purposes invisible from the ground, and thus safe from firearm harassment.

Every time you fly and look out the window, you perform surveillance of a kind, looking down to a landscape revealed while your gaze is invisible to those on the ground. It is a voyeuristic act, to be sure, a privileged vantage point by virtue of excluding others. Los Angeles is one of, if not the most, aerially surveilled cities in the world. I have no qualms about flying over the desert to look and photograph, or over military and industrial facilities. To fly over residential neighborhoods, however, especially in a helicopter, which flies lower and is thus more intimate with what's below than an airplane—that gives me pause. It doesn't stop me from accepting my role as a backseat observer, however.

Once again the helicopter we're flying in is an R-44. The Robinson helos are made in nearby Torrance, and this particular model has just been rated the safest and most popular aircraft in the world. None of which increases my sense of safety from a week ago. We take the doors off for both Mike and Denis, and make sure everything inside the cockpit is securely fastened. It would be severely unfortunate to have a jacket pulled out by the wind and wrapped around the tail rotor. At least I don't spot any seagulls, noting how ironic it was to have been surrounded by them in the desert, but not here on the coast. All I see nearby are ravens, which I normally associate with the desert. I shrug off any sense of foreboding, along with my voyeurism, and climb into the back seat.

Andre lifts us up and rotates north toward neighboring Lakewood. We're all keen to view it from the air, especially given the status of the 1950 photos as rhetorical artifacts in the critique of subdivisions and sprawl.


Given what we now know to be the inextricable connections among the military, flight, and photography, it's perhaps natural that her husband

Beaumont Newhall's military experience as an interpreter of aerial photographs encouraged Nancy Newhall to use aerial images in the exhibition of *This is the American Earth*, and even more specifically to use William Garnett's work as a central visual argument in the book. The first photograph of his that she included is simply titled "Smog," and is taken at a low, oblique angle from above the hills north and east of downtown. The top of City Hall projects upward, but most other buildings are only dimly visible in the depressing murk. On the facing page are three images stacked one above the other, which have since become icons in the antisprawl movement. They are from that commission given him in 1950 by the developers of Lakewood to document the process of mass housing construction. This relatively new way of constructing a community was brought back from the war by William J. Levitt, a custom-home builder who ended up in the Navy's construction arm, the Seabees. The military taught Levitt how to standardize design and materials, procure mass quantities of constructions materials precut to size, and then train crews to erect the same barracks—or homes—over and over again. It also taught him how cost effective it was to bulldoze the ground to suit your needs, versus building around the contours. Other developers soon adopted his economy of scale. Welcome to the suburbs.

The developers of Lakewood built what was then the largest housing project in the world. Their bulldozers displaced four million cubic yards of dirt, and that's the image that follows "Smog," a brutalized abstract of what was once plum orchards, but what now appears to be randomly scarred and anonymous terrain. Heading down the page and in the middle are the foundations for a hundred homes laid out on diagonals to the viewer, a somewhat disorienting perspective. The final and bottom picture, using diagonals running in the opposite direction, shows the finished houses. Taken in low, raking light with dense shadows stretching across the streets, Garnett has emphasized the tight, almost claustrophobic placement of the houses. In none of the three photos is the horizon visible. That is saved for the double-wide spread when you turn the page. Taken from above Lakewood looking back toward downtown, in almost exactly the opposite direction as the first photo, Garnett used one of those magical days in L.A. when you can see from the mountains to the shining sea. And what you see is nothing but houses too innumerable to count. The tall office buildings of downtown are so far away you can hardly be bothered to notice them. The connection between rampant growth, the leveling of the land to suit human purposes, and smog is perfectly implied. The series as a whole makes it looks as if suburban sprawl is war made upon the land, which will become a common trope in urban planning and cultural geography.

Garnett became the photographer who, along with Margaret Bourke-White, did much to make the collective view of the American landscape an aerial one. He was born in 1916, grew up in the L.A. area, and worked first as an architectural photographer, then during the war as a staff person preparing charts for aerial reconnaissance. Garnett had taken his first flight as a kid, a scenic tour in a biplane over the Los Angeles Basin, and had been amazed by the appearance of what he would later discover were alluvial fans, those deposits of erosion upon which much of the city is built. But it was when he was flying back from Long Island after his discharge from the Army, and he was allowed to sit in the navigator's seat for his first cross-country flight, that he grew determined to get his own pilot's license and take photos from the air.

Garnett was awarded a fellowship from the Guggenheim Foundation in 1953, was hired by *Fortune* to take aerial photos alongside Bourke-White, and was given his first solo exhibition by Beaumont Newhall in 1955 at the George Eastman House. Edward Steichen even included him in the famous Family of Man show at the Museum of Modern Art. Garnett's Lakewood photos were recently acquired by the Getty Museum, and when in 2006 I had lunch with Weston Naef, the Getty's curator of photographs, I asked him about the work. Naef, who considers most aerial photography to be more a matter of play than serious art, noted that the prints have an extraordinary tactile quality, but also that Garnett was a pioneer. The Lakewood photos were the only modern aerial photographs that the Getty had in its museum collection.

Andre climbs to one thousand feet and at Denis's suggestion starts traversing Lakewood, first from one direction then another. Mike works his

Linhof and Denis a little digital model. The change in the city below us from the Garnett photos is startling. Gone are the depressingly monotonous rows of bare gray roofs. The houses form a patchwork of color—gray, silver, ochre, and tan roofs interspersed with swimming pools of all shapes and sizes. There's a canopy of trees lining many of the streets, and I'm hard-pressed to find any one house that is identical to another. It's obvious that everyone has been adding on rooms to the original structures for decades. Last year I'd driven through the area with D. J. Waldie, a Lakewood city official and lifelong resident who is also one of the more influential authors in Los Angeles. His book about growing up and living in Lakewood, *Holy Land: A Suburban Memoir*, is a masterpiece of suburban America, and it takes issue with Garnett's photos, five of which Waldie reproduces in the book.

Most of the city's original 17,500 houses are still standing, all built within a single three-year period. Waldie writes that Garnett's "photographs celebrate house frames precise as cells in a hive," and that "seen from above, the grid is beautiful and terrible." Denis, looking down, marvels at the variety now found in the city, precisely the point Waldie made as he showed me how people had modified their houses, customizing them into homes that met their specific needs and desires. Every house has a tree planted in front of it, as required by city ordinance. The eucalyptus, jacaranda, crepe myrtle, and Brazilian pepper trees grew up, and now it's exactly the kind of neighborhood we hold up as an example of "new urbanism," affordable towns where you can walk or take public transportation to most everything you need. Waldie, in fact, has never learned to drive nor ever owned a car.

We inevitably read the Garnett aerial photographs as authoritative appearance, the combination of no-nonsense black-and-white palette coupled with an aerial vantage point insisting to us that they present objective truth. Well, they present one truth, to be sure—but the larger history of Lakewood is complicated and left out of the frame, as only one instant has been captured. The Lakewood tract was first home to the Native American Tongva people, who were displaced by Spanish missionaries, then later in 1790 to retired Spanish soldiers, who grazed cattle and wild horses on the property. During the nineteenth century the enormous rancho was divided and subdivided into increasingly smaller parcels by

Anglo farmers who raised crops such as beans and sugar beets. The agricultural history of the landscaping gave rise to the first template for a Lakewood development, which included a golf course, streets laid out in
graceful Arcadian curves, and plots for large houses, what today would be
marketed as luxury homes for business people commuting into downtown.

Only four houses were built to the original plan, however, and by the
end of the war a modest development of approximately 1,300 smaller
working-class houses existed on the property, an enclave for white workers from the nearby Douglas Aircraft Company in Long Beach and personnel at that city's Army Air Corps base. During the war, Douglas
employed some fifty thousand workers, a figure that would double during the next decade. In 1949 the owners of the 3,500 remaining open acres
sold it to the Lakewood Park Corporation, who promptly adopted Bill
Levitt's economy of scale and mass assembly. During the peak construction period of Lakewood, four thousand workers erected a house every
seven and a half minutes, nailing together a seemingly infinite supply of
precut lumber into standardized forms. But, the developers also included
the world's largest shopping center and sixteen neighborhood retail centers, as well as service roads where children could play out of harm's way
on the larger avenues. There were pre-planned sites for churches, schools,
and parks. As a result, the homes, which ranged in size from 800 to 1100
square feet, literally had people standing in line to buy them. Today, residents are fiercely loyal to a city that is handsome, relatively safe, and a
humane place to live.

As Waldie points out in the official history that he coauthored, *The Lakewood Story*, the city has become "the designated stand-in for all the sins of
suburbia," fixed in the larger American consciousness by Garnett's photos, which were "so memorable that they leave no room for newer, more
generous and realistic images of the city and its residents." Denis points
out as well the inherent division between how the upper class in Los
Angeles literally looked down upon the working class, whose housing
was tied to employment. The upper class could afford to live in custom
built houses built along curving cul-de-sacs in the hills and far from the
sources of industrial pollution, while the workers were able only to afford
standardized housing in the flats.

All three of us are conscious of how the Garnett photos aren't just famous environmental images, but are also now locked into art history as valuable aesthetic objects. And it is true, they are beautiful. Garnett himself didn't like the photographs, as the sprawl and pollution of Los Angeles disgusted him, and by 1958 he was doing well enough to move his family to the vineyards of Napa Valley in the northern part of the state. Like Ansel Adams and Eliot Porter, he preferred his photographs of nature to exclude people and industrial artifacts, to present to audiences what he considered to be an ideal landscape. His acute ability to perceive and isolate striking patterns within a larger field of view was transferred from the urban grid to the sand dunes of Death Valley, the contour plowed fields of Montana, the fractal intricacies of tidelands along the coast. Beaumont Newhall, who knew more about aerial photography than any other curator or critic in America, claimed that Garnett was the only photographer making views from the air based on aesthetics and design; everyone else was doing survey or reconnaissance work. Obviously Weston Naef agrees.

Once we're done peering down at Lakewood, Mike asks our pilot to bear north and over the Atlantic Richfield Oil Refinery, which abuts the city. Andre keys open his mike. "We're coming on station over those big tanks," he reports to the tower. We're over the 91 Freeway and in between the complex and invisible approach patterns for two airports, a very busy place. Los Angeles has thirty-five airports and airfields, more than any other urban area in the country, and the Long Beach Municipal Airport is one of the busier general aviation fields in the world. Andre stays in constant contact with the controller so traffic control knows where we are. We orbit over the storage tanks and the refinery, and I'm amazed that we're allowed to roam freely over an industrial site, given the proximity of residential neighborhoods with the potential for an accident or act of terrorism.

When the guys are finished photographing, we follow 91 west and across a power line corridor and then over the Los Angeles River. Under the high tension wires are rows and rows of plantings, a commercial nursery leasing the space, which is at a premium. We had no idea people had exploited these interstitial spaces for commerce, much less for growing flowers. Driving by on the freeway you might miss the low rows of culti-vation, you'd be so busy keeping track of your position relative to the rest of the traffic. On the far bank of the enormous concrete gutter that is the river, Freeway 19 meets the 710 going north, an interchange with what, quickly counted, looks like nineteen lanes of traffic on four levels so visu-ally complicated that only an engineer could love its knotted coils. But from the air, the openness of the ground and patterned planting next to it draws your attention immediately, a sharp contrast to the built environ-ment, a garden in the machine.

The river, which begins in the northern reaches of the San Fernando Valley, empties into San Pedro Bay to our south. Mike's first major aerial series over the city, *Los Angeles 02.12.04*, followed the river from source to outflow about a thousand feet above the watercourse. The black-and-white photos are seldom level with the horizon, low obliques that often point directly into the sun and haze, and are virtually uninhabited by any-thing but cars and trucks. You walk away from them with two impres-sions: that L.A. is a relentless horizontal grid created by concrete and asphalt, and that, despite the towers of downtown, the only real vertical relief is offered by the Hollywood Hills. None of this seems surprising when you think about it, but visually the extent and seemingly endless variation of the grid still has the power to astound. You can hardly pull yourself away, trying to figure out where you are in the maze.

That's a common attraction to aerial photos, trying to find your place, if not your house, then at least familiar landmarks. But Mike—by empha-sizing how little we can discern from the air, by making what by Garnett's standards would be supremely ugly pictures that obscure much of the city, versus making it more obvious—opens to us aspects of the city we fil-ter out when living here. First, he does what any person coming into new territory would do—he follows the river, the water, because that's what everything will have originally been built around. That was certainly true of Los Angeles, which started out as a Spanish pueblo built on the east bank of the river. But no Angeleno driving around today uses the river as a landmark; its presence is almost invisible from the ground until you're driving over its concrete channel. So the route that Mike followed in the photographs is not obvious even to most local residents.

Second, he was taking pictures after a three-day Santa Ana wind event that had completely cleared out the smog. The photos disappear into a haze that we, as locals, automatically assume is pollution, but it's not. It's mostly just the ocean-blown moisture and particulate-induced airlight. We forget that the light here that is so prized by filmmakers for its evenly diffused quality was that way prior to the rise of the automobile. And third, Mike's photos are most often decentered. Yes, the Hollywood sign on Mount Lee may appear in a picture, but it's not the focus of the image; it's off to one side. Anyone driving around L.A. knows that the sign, unlike the river, is a visible landmark.

Today, as Mike leans out of the helicopter door frame, as usual pushing himself to the limits of the harness, he's doing the same thing, albeit in color film. Once past the interchange, we cross over what remains of Compton Creek, a concrete ditch that dead ends into stagnant water, a slick of chemicals floating on its surface. The Del Amo Mobile Estates, a trailer park, sits next to a warehouse district and a circular dead space. I mean wilted vegetation, cracked pavement, trees gone feral, and no one in it. It looks like an abandoned park, but given the premium on space in the urban area, it has to be something else. Later we discover it used to be a car park, the Compton Auto Plaza—not a place where you take your kids to play, but a sculpted retail environment set around a central circle of trees. The site is empty because a big-box retail store is slated to be built on it.

"There's a steep social gradient here," Denis points out, as we transition from manufactured housing to a brand new subdivision with two-story homes. As we head farther west toward the Wilmington refinery, we can smell the hydrocarbons. And then there's a huge, European-themed apartment complex with red tile roofs and curving streets next to an abandoned housing development built by the Port of Los Angeles. The concatenation of industrial facilities and old housing with golf courses and upscale developments is beyond all our expectations. What produces such juxtapositions is not the flow of water but money, a complexly braided flux of tax credits and shelters, investment strategies and bankruptcy.

As we head south toward the conjoined ports of Long Beach and Los Angeles, the busiest such complex in North America, the landscape is comprised of rank upon rank of new cars from Japan and Korea, most of the roofs of which seem to be red, white, and blue, surely not a coincidence, given the closeness of the American identity to automobility. Nissan, Honda, and Toyota all have major corporate, design, and shipping facilities here, located within blocks of the major refineries that produce most of the gasoline for Southern California. The money may be invisible from this altitude, but it's physically manifested in trade goods, the scale of which is, again, revealed from an elevated vantage point. We're cruising over shipping containers, tens of thousands of shipping containers stacked three, four, even five high. "There's the trade deficit right there," observes Denis. He's uttering more than a metaphor.

The shipping container, at first simply the trailer of a semi truck with the wheels removed, was invented in the late 1940s as a way of getting around the bottleneck caused by the loading and unloading of cargo ships, a process that could keep them tied to a dock for as a long a time as they spent sailing. As the ships got larger, the transfer of goods was taking increasingly longer, and it became inevitable that the process would have to be sped up by standardizing it. It's only logical that the answer was developed at the same time as tract housing was invented after World War II, the conflict having seriously modernized the study of logistics. The first container ship was a reconfigured military tanker that carried fifty-eight containers out of Newark, New Jersey to Houston, Texas. Nowadays the capacity of the container ships is measured in TEUs, or Twenty-Foot Equivalent Units. By 2002 the ports here were handling more than 10,629,902 TEUs annually, and more than 22,000 inbound ships. That's a quarter of all the containers entering America. The vessels now carry up to 14,000 TEUs each, and are so large they don't fit through the locks of the Panama Canal. Ships now on the drawing boards—well, computer screens—will carry more than 18,000 TEUs, and won't even fit through the Suez Canal, which has no locks to squeeze through, but simply isn't deep enough for their passage.

That's all well and good, but we've been running a huge trade deficit with Asia. More stuff arrives here than departs, and it's cheaper to leave the containers sitting around than to ship them back to their points of origin. And that's what we're looking at, the physical evidence of an economic imbalance. Needless to say, the genius of the free marketplace is already at work. You can buy a used forty-foot container on eBay these days for around $2,000, and they are well-built enough that they're being used for everything from security huts and concession stands to garages, and even as additional rooms on houses in Santa Monica.

As I'm scribbling we pass over a newly created peninsula for expansion of container business, a square of raw earth pushed out into the water, the grid extending itself out from the land. The coastline here is not so much a cleanly delineated edge of the continent as a fuzzy zone of interpenetration, the harbors being dredged and deepened even as the land is pushed outward. To our left in San Pedro Bay sit four artificial islands constructed in the mid-1960s, shallow water drilling operations hidden behind walls covered in bright abstract designs, and adorned with palm trees, oleanders, and waterfalls. The landscaping was created by Joseph Linesch, who used his work on theme parks such as Disneyland as inspiration for what became known as "fantasy islands." His $10 million camouflage job disguises 180-foot-high drilling rigs and 1,100 wells that produce 32,000 barrels of oil a day. When they run dry several decades hence, the four ten-acre islands will revert to city property.

The islands, the nearest of which is only two thousand feet from the beach, were built before the infamous blowout of a well off Santa Barbara in 1969. The first offshore drilling in the world occurred near Santa Barbara in 1897, but the accident, which coated thirty miles of beach with black goo, produced a severe community backlash with national reverberations, one that led to a moratorium of offshore exploration and development that's still in place, despite the rising cost of oil. These Long Beach rigs hit their peak pumping in 1995 at 72 million barrels; by 2004 they were down to 24 million barrels and still declining. I think about Kennecott building its Daybreak community on top of mine tailings, and know that here, despite the deep contamination of the soils, real estate money will ooze into available land when the oil runs out.

After we've passed over the breakwater, Andre takes us down to only a hundred feet above the sea at 90 knots, the nose of the helo canted sharply downward so the rotors get enough forward bite to produce the speed. At just over 100 mph, the waves rush at us as if a movie were unreeling on a screen, and sure enough, what we're experiencing is often filmed as a common cinematic device to make the audience feel as if it is zooming in toward the scene of action: an island, a vessel, or the shore of a city. A seal rests on its back and watches us roar overhead with no apparent concern, evidence of how settled these waters have become around the drilling and pumping facilities. The platforms here aren't huge. A derrick, a helo pad, a crane, a jetty. The sea bottom isn't that deep, and they aren't that far off the shore, so the oil is just piped to the nearby onshore refineries. The really giant rigs in places such as the North Sea have storage facilities that hold the oil until it can be pumped into tankers. We circle one of the seven rigs out here, the "Eva" platform. Seals lounge everywhere on buoys, a surface indication of the small but rich ecosystem of fish and mollusks that's grown up around the underwater pilings. The colonization of territory around the industrial sites isn't accomplished only by humans in trailer parks.

We head back to the east of Long Beach to fly over the next community down the coast, Seal Beach, which is its own study in contrasts. Moving inland, the sequence is as follows: Sandy beach, scrim of vegetation cut through by footpaths from each house situated slightly above the beach; street, red-roofed luxury condos and apartments, swimming pools, more residential units, marina. The marina, one of those artificial interpenetrations of the coastline, is a complicated array of houses built around cul-de-sacs which back onto canals and slips with more boats than I've ever seen in one place. The marinas in the area host 3,400 individual slips, and most of them are filled. But immediately south of this high-density aquatic development is empty wetland, 920 acres of salt marsh preserved as a national wildlife refuge. It's part of the Naval Weapons Station, a 5,256-acre facility that stores and loads missiles, torpedoes, and conventional munitions onto naval vessels from its hundred-foot-long wharf. Virtually the only undeveloped coastline in Southern California is military, places such as here, the U.S. Marine base at Camp Pendleton to the south, and the Air Force space defense facility at Vandenberg to the north of Malibu.

We're close to restricted airspace, and veer north, back toward a large grouping of long brown and gray roofs covering—well, at first I think they're barracks, or abandoned mass housing units. It turns out to be Leisure World, the country's first planned retirement community of its size. Built in 1960, the community covers a square mile and holds nine thousand people in what constitutes half of Seal Beach's entire number of households. A few parked cars, no one visible walking around, windows invisible under eaves—it all has the feeling of a well-maintained ghost town sandwiched in between the munitions and yet another oil refinery.

The last place we fly over is Signal Hill, which rises a stunning 365 feet above sea level just southwest of the airport. Signal Hill was explored for oil in 1921, although it had already been subdivided and sold for residential development. When the first gusher came in, property owners promptly gave up their plans to build houses and threw up oil rigs as fast as they could, more than three hundred of them, some packed so closely together that their wooden legs were erected within each other. For years the enspined rise was known as Porcupine Hill, and it became one of the most productive oil fields of the twentieth century. The original and more enduring name of Signal Hill derived from its use by Tongva Indians as a height upon which to build signal fires observable across the basin and over to Catalina Island, some twenty-six miles out to sea. It still hosts radio transmission towers, which we carefully avoid.

As we circle around what is now an upscale housing development with contoured streets and terra cotta roofs, we spot pumps chugging slowly up and down in backyards. Some eleven thousand people live within the 2.2 square miles of this elevated enclave that is completely surrounded by Long Beach, most of them in apartment buildings, but a few in homes that cost upwards of $2.2 million. Why would anyone pay that much to live under the flight path of an airport, breathing in dense hydrocarbons while contemplating the proximity of aging and increasingly accident-prone refineries next to a munitions dump, not to mention the planned liquid natural gas terminal for the harbor—which has the explosive potential of a small nuclear device—and all of it in earthquake country prone to typhoons and small tornadoes blowing in off the Pacific Ocean? It boggles the mind. Maybe it makes sense for a retirement community, because what's to lose? But to raise a family here? It's also true that you wouldn't think so much about the absurd contrasts if you weren't in an aircraft and thus able to take in all of it at a single glance.

Andre gets us back to the Long Beach airport at 5:20 P.M. without much to-do. Mike's taken only one magazine of pictures. The flat, murky light, a kind of dim aerial dome over an urban space, has been in complete contrast with the white open pages of the Nevada and Utah salt flats, and the steep verticals offered by both the Bingham Pit and the Garfield Stack. But L.A. has long been represented from the air, and not just from airplanes. The earliest known example of an elevated view of which I'm aware is a drawing of Los Angeles from 1854. Made from a hill across the river and looking east, it shows slightly to the left of center the original plaza of *El Pueblo de Nuestra Señora Reina de los Ángeles sobre El Rio Porciuncula*, which is Spanish for The Town of Our Lady Queen of the Angels on the Porciuncula River. Stretching to the right and south are the houses that are beginning to grid off the ground. In 1887, when the town was about twenty thousand persons, a photographer managed to make a picture of the city from a balloon flying east of the city. By then the grid stretched across hundreds of acres, still oriented to the winding river channel and thus offset various degrees from the strict cardinal orientation we use in the twentieth century.

# 6. The Angels of Mulholland Drive

*Altitude is the muse of enlightenment. We seek high ground in order to gain perspective on our environment and our lives, to steal away from the clamour of the streets and low places and reflect on our being-in-the-world. Elevation extends our vision, literally and figuratively. The complexity of life is reduced to utopian simplicity, a living diorama as benign as a child's train layout. . . . In almost every area of cultural production—literature, philosophy, urban design, even politics—elevation is a symbol of knowledge and power.*

—Thomas J. Campanella

**The classic elevated view of Los Angeles is that of a vast horizontal** grid, sometimes made from the air, but sometimes from the Hollywood Hills, specifically from Mulholland Drive at night. It's the glowing grid that stretches endlessly into the distance as seen from the crest of the hills, a view that photographers and painters have had to cope with ever since the road opened in 1924. It's a view that fits within the long tradition of bird's-eye views made of cities since the beginning of the Renaissance in Italy, and yet it's more metaphor than map. Mulholland runs fifty-five miles east from the Pacific Ocean along the ridge of the Santa Monica Mountains, crosses Interstate 405 at Sepulveda Pass, winds the length of the Hollywood Hills, and ends at the Cahuenga Pass just northeast of the Hollywood Bowl.

It's this drive that has provided the archetypal views of L.A. as a net dense with connectivity, and brought us to compare the grid to a machine, a computer, a mind. And it is the image that first comes to my own mind when I think Los Angeles. Not the beaches, the skyscrapers downtown, Hollywood and Vine, Beverly Hills, or my great-grandmother's former house near Wilshire Boulevard where I said my first word at three ("car").

It's early rush hour on a winter afternoon, and I've just left the Getty Center in the hills above Santa Monica for the day. Usually I get on the 405 and head north over Sepulveda Pass, merge right onto the 101 and then pick up the 134 to get to Burbank and home. But today, because I was delayed by an event, I left late. The intersection of the 405 and the 101 is the busiest freeway interchange in North America, and one avoids it at all costs during "rush hour," which lasts from about 6:00 to 10:00 in the morning and from 4:00 to 7:00 in the evening. That's a full third of a 24-hour day, in case you're counting. So I have, instead, opted to saunter home along Mulholland. The meandering two-lane road will take me less time than on the freeways, about forty minutes compared to an hour and a quarter, and it will be far more pleasant. And it's a fitting way to end this day, which has been all about cities and our aerial views of them.

The event causing my tardiness was somewhat of my own making. For the last two years I've been working at the Getty on *Aereality*, and today was a major waypoint along my peregrinations, a symposium in which Denis, Mike, and I, along with our colleague from Italy, Lucia Nuti, talked about "The God's Eye View." Our premise was that, just as people typically look heavenward as a way to approach the divine, so they attempt to bring themselves closer to understanding and perhaps emulating the divine by elevating their own vantage point to look down on the Earth both imaginatively and physically. Nuti started off by tracing the development of aerial views in Western culture from the Renaissance to the twentieth century. She noted that the concept of a totalizing vision from the sky, as an expression of divine faculties and a superior form of knowledge, was rooted in the Greek world, where it echoed in myths and poems. When the same concept emerged again in Renaissance Italy, it was translated into an image, the perspective plan of the city, a view invented to fulfill the ancient dream and overcome the limits of human vision.

I had spoken next about the possible cognitive roots of aerial perception, first showing the image of the Çatalhöyük mural and ending with recent work done over Los Angeles by Mike's fellow Bay Area photographer David Maisel. My point was that I had started off my research thinking that the human ability to imagine oneself above the Earth and looking down was an exceptional and rare gift, and that commercial air travel after World War II allowed the public at-large to assume such a vantage point in reality. But, the more I studied aerial images throughout history, the more convinced I was becoming that the aerial perspective was actually a normative one, something we all did all the time without necessarily being conscious of it.

After lunch, Mike led off by showing the work he'd made over Bingham and Los Angeles flights. He spoke to one of the central paradoxes of the Olympian aerial perspective, experienced today in varying degrees by everyone from astronauts to tourists: "the vaster the view and sweep of knowledge revealed, the smaller the seer becomes in relation to that view." Even though the viewer gains an enhanced ocular perception, he or she is always constrained by personal physical scale. That is, the "sky voyager is at once both aggrandized and diminished by vertical height over the ground." He addressed what it was like to work physically *in extremis* trying to master mind, eye, and recalcitrant machines to move beyond mapping and create art.

Denis wrapped up things with a talk about "God, the Globe and Google Earth." His historical images demonstrated how monotheisms "relocate divinity from the landscapes and spaces of the material cosmos to a transcendental point beyond the earth's surface, with God as a creator contemplating the wonder of 'His' creation." He reminded us that humans, according to Judeo-Christianity, were made in God's image, but placed at the midpoint of creation between angels and animals. As a result, "humans have consistently imagined what it might be like to attain God's perspective over the globe to whose surface they are bound." After running through a series of images showing how this desire has been expressed in Western art, Denis dissected how that history framed our response to the first "whole earth" photographs taken by astronauts. He finished off the day by using Google Earth to zoom down to the planet's surface as if it were a realization of our ancient dream of assuming the eye of God.

So there you have it, the traditional humanist's take on aereality: It's part and parcel of our search for the Divine, but also based in our physical nature. Which is to say, our ability to see stems from both nature and nurture, or nature and culture, if you prefer. The ideas I outlined about the cognitive roots of aerial perception were preliminary, and I'll come back to them later in more detail. But for now, it's time to just take a look around from a high place. To see how the Eye of God looks upon the City of Angels.

Mulholland Drive was conceived of and built by many of the same people responsible for the Los Angeles Aqueduct, that visionary, if supremely dubious public works project that sucked water from the Owens River Valley 233 miles south to the San Fernando Valley. The aqueduct was completed in 1913, and around the same time engineer William Mulholland was muttering about a road atop the Hollywood Hills. As Mulholland was overseeing the construction of the aqueduct, the L.A. city engineer for street design, H. Z. Osborn Jr., began arguing that the basin needed a grid of arterial boulevards to accommodate urban growth. Political opposition by property owners was stiff, and if wasn't until 1924 that a team of consultants was able to present a plan that the nonprofit Traffic Commission could get passed. The 1920s saw the population of L.A. triple, the number of cars boom, and relief from traffic became more palatable to the populace. Not much has changed; one of the reasons I'm avoiding the 405/101 interchange today is that the city is widening it to accommodate a population of people and cars that's still growing. The boulevards that Osborn had envisioned were built, but Mulholland Highway was destined never to become part of the transportation network as such. Instead, it quickly became known as a scenic drive from which you could apprehend the horizontal nature of the city.

Developers have tried since then to widen what is now called Mulholland Drive, interested particularly in the part I'm driving today, the 11.7 miles south from Sepulveda Pass down to where the 101 passes through Cahuenga Pass. This is the stretch that connects the tops of all the canyons made famous in movies and songs—Beverly Glen, Coldwater, and Laurel among them—and every now and then someone would dream up a scheme to build a hotel atop the hills. But since the 1970s the road has been protected as a designated scenic parkway. The drive has been favored by lovers stopping to romance, motorcyclists racing at insane speeds, movie stars driving to their houses, and wayward cultural geographers perching to contemplate the city. Most of my drive for the first ten miles hugs the northern edge and offers fine views of the San Fernando Valley, and I pull over frequently to let the Mercedes, Porsches, Lexuses, and even a Maserati pass me by. Many of the houses on either side sit behind fences, gates, hedges, and lascivious tangles of red bougainvillea spilling over walls. The houses themselves range from shambling three-story wooden-frame rentals on stilts to Spanish mansions, Mediterranean villas, and French chateaux.

From up here the straight and broad boulevards of the Valley—legendary streets such as Sepulveda and Van Nuys—run north across its 375-square-mile floor upon which 1.7 million people, myself included, live. The Valley is big, but it's bounded; you can see how the Santa Susanna and Verdugo mountains enclose it. At about the ten-mile mark along Mulholland, however, with the road now crossing to the southern side of the ridge, the view opens up to the L.A. Basin and downtown. And suddenly I'm gazing out over what seems to be that endless urban grid. It just disappears into the haze and over the horizon. You can't help but pull over and stare at it.

As I said before, think of Los Angeles, and most likely the first image that comes to mind is this grid before me, but at night. At first it was pictured from Mulholland, and then increasingly after World War II from airplanes. The night views were more typically presented in film shots or in paintings, given the technical difficulties of making still photos with low light in vibrating aircraft, or they were presented to us in literature. I don't usually quote the late French cultural theorist Jean Baudrillard, especially about matters of American geography, which he so often gets spectacularly wrong, but he was a photographer and he nailed this view on the head in his 1986 book *America*:

There is nothing to match flying over Los Angeles by night. A sort of luminous, geometric, incandescent immensity, stretching as far as the

eye can see, bursting out from the cracks in the clouds. Only Hierony-
mus Bosch's hell can match this inferno effect. The muted fluorescence
of all the diagonals: Willshire [sic], Lincoln, Sunset, Santa Monica . . .
You will never have encountered anything that stretches as far as this
before. Even the sea cannot match it, since it is not divided up geometrically. The irregular, scattered flickering of European cities does not
produce the same parallel lines, the same vanishing points, the same
aerial perspectives either. They are medieval cities. This one condenses
by night the entire future geometry of the networks of human relations,
gleaming in their abstraction, luminous in their extension, astral in their
reproduction to infinity. Mulholland Drive by night is an extraterres-
trial's vantage-point on earth, or conversely, an earth-dweller's vantage-point on the Galactic metropolis.

Of the 245 photographs collected in the recent photo anthology *Looking
at Los Angeles*, almost 20 percent of them are from either elevated vantage
points looking down into the city from the hills, rooftops, or upper floors of
office buildings, or outright aerials. Because this is a book that purports to
be a photographic tribute to and history of the city, it's worth examining the
opening images, which start with the cover by Florian Maier-Aichen, a
young German artist who splits his time between Cologne and L.A. It's a
2002 view of the nighttime grid taken from the summit of Mount Wilson
some fifteen miles away and five thousand feet above downtown. The city
is so dense that the lights melt together in a milky haze enhanced by the
photographer and that obscures details beyond the first few large blocks
delineated by the large boulevards and freeways. The format of the book is
horizontal, and the photograph reinforces that Los Angeles is likewise a
densely patterned carpet versus a forest of buildings like Manhattan.

The title page of the book is one of the obliques taken by Mike. In the
immediate foreground is a parking lot, the rest of the city intersecting axes
of boulevards. Two parallel rows of approximately seventy evergreens
proceeding from left to right planted along railroad tracks. Within only a
few blocks, the city disappears into a white haze. Several hundred cars,
either parked or in motion, are visible. In the original print and through a
magnifying glass—but not visible in the book—are two *rara avis*, two

pedestrians on a sidewalk. The city is flat, paved, mechanized. Turn the page and there's a night shot of the San Fernando grid, North Hollywood as pictured from up here on Mulholland Drive by Doug Hall. Turn the page and there's one of Ed Ruscha's classic parking lots, a vertical aerial from his *Thirty-four Parking Lots* done in 1967. The white diagonal lines demarking the empty rows and individual spaces are offset by the black oil spots left by the cars. Turn the page and there's the color photo *Overview, Angels in Fall* by Karin Apollonia Mueller, a low oblique of the daytime city made from a dirt shoulder above the grid which, like Mike's photo, simply fades to white. Turn the page and there's a 1945 aerial by Will Connell of *Suburban Homes*, one-story single-family dwellings for GIs returning from the war stretching out of sight. The first line of the foreword on the facing page by editor Marla Hamburg Kennedy reads: "This book is about our love of Los Angeles." Apparently we are enamored of the grid from above.

But, then, this is true of almost any human settlement from Çatalhöyük forward. We always want to see the pattern of which our place is a piece, as well as the places we have visited. The first known attempt at making topographically accurate views of cities, at least in the West, are in a remarkable travel book by Bernhard von Breydenbach, who lived from 1440 to 1497. *Peregrinatio in Terram Sanctam*, or "Journey to the Holy Land," was published in 1486. The *Peregrinatio* was the first illustrated travel book and the first book to credit an illustrator on the title page. It was widely reproduced via woodblocks across Europe as a guide to travelers, became the best-selling book of its time, and contemporary books reusing the title have continued to be published worldwide.

Breydenbach hired Erhard Reuwich, whom he called "a skilled artist," to document the journey, and they sailed to Iraklion, Modoni, Rhodes, Venice, Corfu, and Parenzo before reaching the Holy Land. At each city Reuwich made sketches of elevated oblique views that captured the overall layout of the towns with their major buildings. The views were made as a kind of running profile from an imaginary traverse in the air, more like the profiles made by sailors of coastlines than an aerial map made from a static viewpoint.

Although a few European townscapes were made prior to Breydenbach's journey, the publication of his book created a huge appetite amongst Europeans for aerial views of their own cities. The impetus may have been twofold. One, as cities were getting larger, people could no longer see out to the edge of town just by looking down a street. It was becoming more difficult to place yourself in the world. And, two, as competition among cities for trade increased, it benefited the merchants to promote a more sophisticated civic identity. The European capital of mapmaking and global trade at the time was Venice, and around 1497 the publisher Anton Kolb commissioned the Venetian painter and engraver Jacopo de' Barbari to create an aerial perspective of the city. Unlike Reuwich's views, which offered relatively simple outlines of major buildings, this would be a fantastically detailed accounting of each building and street in the city. It took the artist and a team of workers climbing bell towers three years to assemble the picture, which Lucia Nuti argues is more a painting than a map. But it was so accurate that historians today still refer to it as a baseline reference document.

Throughout the sixteenth century the techniques for measuring height and distance would steadily improve through the efforts of military officers laying siege to various cities, and who needed to improve constantly the accuracy of their bombardments, tunnel excavations, and breaching tools. By 1570 Abraham Ortelius of Antwerp could assemble the best maps in the world for the first edition of the world's first modern atlas, the *Theatrum Orbis Terrarum*. Included were bird's-eye city views. Two years later his friend and colleague Georg Braun would begin to publish a compendium of nothing but city maps and views into what by 1617 would be the six-volume *Civitatas Orbis Terrarum*. Braun gathered together more than a hundred cartographers, painters, and surveyors in order to create the atlas of 546 cities that ranged from Moscow in the north to Cairo in the south. Nuti points out that the bird's-eye views of cities published by Braun were so seductive that travelers to Italy in the sixteenth and seventeenth centuries climbed bell towers in the towns pictured to find the same vantage points assumed by the artists. Invariably, they were unsuccessful. The artists had, indeed, used those same bell towers to inform their representations, but the vantage points of their aerial views were entirely imaginary.

In America, elevated panoramic views were being made as early as 1719 of seaports such as Boston, but true bird's-eye views didn't really become popular until the nineteenth century, when something on the order of 5,000 were made from coast to coast of at least 2,400 separate locations, according to John Rep's study of lithographs made during that century. The civic pride of Renaissance Italians had nothing on the manifest ambitions of western expansionists, and as Rep points out in *Bird's Eye Views: Historic Lithographs of North American Cities*, the American practice was uniquely democratic: the views weren't made of just major trading centers, but also small towns such as Moscow, Idaho. Artists would walk the town streets, sketch the facades of every structure, then do a perspectival drawing of the city grid and fill in the buildings. Some of the itinerant artists were so skilled that they could do a dozen or more such views a year. The aerial views became the most popular lithographs of the century, and were used by land speculators to promote development, then by residents to orient themselves. And Los Angeles, in a land boom on the opposite coast from the older and richer cities of New York and Boston, had everything to prove and land to sell. How appropriate that the City of Angels, then, would be represented iconically by the God's Eye view of the world

In 1857 the partnership of Kuchel & Dresel made a low oblique of the pueblo's Spanish adobe buildings situated around the central plazas with snow-covered mountains in the background. It's a somewhat intimate view, made as if from a third-story window. By 1871 Augustus Koch found it necessary to position his vantage point as a high oblique in order to show how the growing town was surrounded by ranchos and groves of trees. With hindsight the view seems both bucolic and a come-hither—as in "Here's plenty of open land with nothing but orchards on it, come build." Then in 1877 Eli S. Grover produced a large panorama as if drawn from hills in Griffith Park looking out toward the ocean to south and west. It's patently clear in this view that the primary resource of the city is the basin itself, flat land over which to spread for hundreds of square miles.

A national depression in 1893 slowed the land boom and the popularity of bird's-eye views slid as well. In the early twentieth century the country bounced back and cities began growing so fast that the views became quickly obsolete. For one thing, transportation was increasingly mechanized, first through trolley cars and light rail lines, then with the automobile. In a city such as Los Angeles, the availability of land coupled with the spread of oil fields and purchase of automobiles meant a geographical explosion. Aerial photography arrived just in time to address the problem, and the traditional viewmakers began to complain that aerial photographers could document in a single day what they would take months to accomplish. A bird's-eye photograph of the city made in 1902 by J. W. Austin is a good example, a wide-angle panorama looking north that is filled with a grid from edge to edge. The view that I'm looking at from the front seat of my car perched on Mulholland Drive was already outlined by then.

After World War I there was a surplus of aircraft and pilots, and both found their way into commercial aereality. Passenger service was a distinct possibility, the postal service was increasing airborne delivery, and a number of aerial photography firms were formed. Among the more prominent were Spence Air Photos and Fairchild Aerial Surveys, many negatives from which are archived at UCLA, where Denis has been working through them. Sherman Fairchild invented an improved aerial camera during World War I, started making aerial timber surveys in Canada after the conflict, and then pieced together the first aerial map of New York City, a landmark in cartography. He went on to supply both cameras and aircraft to the military in the next war. While Fairchild concentrated mostly on surveying the eastern United States, Robert Spence worked over southern California in depth starting in 1920 for more than fifty years, interrupted only while he served during World War II taking aerials for the military over Burma. Both Fairchild and Spence photos were used by developers to produce surveys for real estate deals, and in fact their photographs will still occasionally show up in court cases as evidence.

Aerial views, like any other system of representation, are hardly neutral documents, however. While the Spence photos over Los Angeles might at first be read as clear of any overt agenda, their use by Charles Owens at the *Los Angeles Times* displays how they were harnessed to the ambitions of the city's leaders. Denis points out that what Spence might have taken to be scenically pleasing views of the city also served to help form a cohesive regional identity for L.A. as a metropolis. Owens was a young artist hired on at the paper in 1919. He worked there until 1952 and specialized in the aerial views of the city that Harry Chandler, owner of the paper and a fervent civic booster, needed to promote development. Using photos by Spence and flying over the city himself, Owens created high, oblique aerial depictions that were both map and picture, document and advertising. Whether it was a page of the newspaper that showed the driving routes around a twenty-mile radius from downtown, or the route of the pipeline from Owens Valley, Owens (no relation) helped promote the Chandler agenda. His imagery, because it was a direct descendant of the nineteenth-century bird's-eye lithographs, was easily assimilated by the public, and his newspaper maps were so popular they were reprinted as brochures for people to use in their automobiles. Los Angeles was now so large that you couldn't navigate around it without an aerial view on the seat next to you.

You can't drive Mulholland without stopping at one of the overlooks and contemplating the city from one side of the ridge or the other, and I park to take a look out over the Los Angeles Basin. I know that my view of the world was altered as a child by looking at the aerial views that Owens and others were making of the world. When I think about the contemporary artists who have painted and photographed L.A. from the air, I can see how they have moved from trying to explain the world to us to making it strange again, a way of seducing viewers into paying more attention to where they live. It's a sad fact that most people will spend only a few seconds in a museum staring at a famous painting by Rembrandt or Van Gogh or Picasso, but when confronted with an aerial photograph of their city will spend minutes before it, locating familiar landmarks. We are so driven by the need to locate ourselves, a survival mechanism hardwired to our core functions, that we are transfixed by the view from above, whether it's in person, as here on Mulholland, or in one of Mike Light's photographs.

# 7. Surveillance

*Every aerial image is essentially a future with a vector for breaking into flight.*

—Gaston Bachelard

**It's a natural enough progression that's been followed around the** world, from mapping a terrain to picturing a territory, moving from chart to art, from cartographic representation through colonization and into symbolic landscapes, and Los Angeles is no different. The aerial views of it have changed from the conventional bird's-eye views promulgated by developers and their allies to those made by contemporary artists seeking to use its urban geography as a metaphor of our times. Denis Cosgrove argues that L.A. is a modern city, its density horizontal and not vertical, as in older cities, and that the aerial synoptic view helped make it so. While parked up here on Mulholland and looking out over that spread, I have specifically in mind contemporary artists such as James Doolin, Ed Ruscha, Peter Alexander, and David Maisel. Each of them has something to add to our understanding of how we see the L.A. grid from above.

One of the aspects of the aerial that Denis and I have been discussing is how World War II increased public awareness of aerial views. The European and Pacific theaters of the global conflict were so vast, and thus hard to visualize, that artists such as Charles Owens were forced to take their bird's-eye perspectives and apply them to entire quadrants of the planet so that readers could follow the strategies and battles of the war. And for

the first time much of a war was being conducted in the air via reconnaissance planes, bombers, and fighters. Aerial photography was no longer
just looking downward, it was also capturing air-to-air images of dogfights. James Doolin was born in Connecticut in 1932, and as a child during grade school was so entranced by images of dogfights that he began
to draw them during class. Patricia Hickson notes in her essay for an exhibition given of the artist in 2001, the year before he died, that as Doolin's
aerial compositions became more complicated he taught himself perspective in order to portray the three dimensions of aerial combat on a two-
dimensional piece of paper. All that while he was ten years old. During
high school his interests shifted into comic strips depicting crime fighters,
and he expressed a desire to become a professional artist.

His early efforts paid off, and Doolin won a scholarship to attend
Philadelphia's University of the Arts where he took courses in both commercial and studio art. He was drafted into the Army in 1955, trained as a
stenographer, and was stationed in Europe where he had a chance to visit
various countries and their museums, an informal continuation of his arts
training. He was discharged in 1957 and for the next several years took
turns working as a commercial artist, traveling to Europe, and making studio art whenever he had the time. He called his paintings "Artificial Landscapes," essentially geometrical fields of color based on photographs he
was taking of the built environment. Doolin had married an Australian
woman he'd met while in Greece, and in 1965 they moved to her home-
town of Melbourne with their two children. Doolin taught at a local art
trade school, and quickly established a reputation as a painter of note with
his abstracts. In 1967 he was accepted to the Masters of Fine Arts program
at UCLA, and he returned with his wife to settle in Los Angeles. Although
Doolin continued to paint his abstracts and show them in Australia, his
studies with the likes of Richard Diebenkorn began to shift his work
toward the representation of actual landscapes as an illusionistic space. In
his graduation show he exhibited both kinds of paintings, a deliberate
statement about where he'd been and where he was going.

In 1973, while Doolin was teaching at UCLA, he began research in
preparation for painting his epic *Shopping Mall*, a 90 x 90 inch canvas that
took four years to complete. His premise was to look down upon an

urban intersection and depict with a vertical impartiality all the variety of life on the street. Doolin selected the busy crossing of Third Street and Arizona Avenue in Santa Monica, right at the front entrance of the mall. He studied maps and architectural drawings, made axonometric studies of the buildings from their rooftops, and decided to compose the picture with the streets crossing at a diagonal in a giant $X$. After making numerous studies on paper, he hired a helicopter from which to make a final check of the shadows during the selected date and time to be depicted, 4:36 on a Saturday afternoon. Diebenkorn, a painter who used aerial views for inspiration in his own work, accompanied him.

Hickson describes the results: "In effect, Doolin created a meticulously reconstructed space—another artificial landscape—that begged to be looked at. From a distance, it is quite abstract. Upon approach it reveals an omnipotent, godlike perspective. Up close, the viewer falls into the composition, and the infinite narratives of its 365 figures unfold. Ordinary events begin to appear extraordinary." The axonometric perspective Doolin used is also known as planimetric because it is based on taking a simple plan of the ground and tilting it up 45°. The technique preserves all the vertical lines as parallel to one another and gives the impression of all objects, be they people or buildings, as having equal emphasis. "All things equal in the eye of God" is a phrase that springs to mind.

Doolin couldn't have even envisioned, much less carried out such a work without his training as a commercial artist, and that background served him well throughout the successive stages of his career. His marriage falling apart, he spent the next three-year stint by himself in the Mojave Desert, where he made imaginary aerial views of lonely vehicles driving through the nighttime desert, their headlights throwing out long cones of light across mural-sized images of an almost alien landscape. When he returned to a studio in L.A.'s industrial district, he brought back with him a harsher palette, a flatter field of effect, and a widened sense of perspective honed in the open spaces of the desert that was more fish eye than bird's eye. From the mid-1980s onward he concentrated on painting the downtown and freeways. Gone were the axonometric perspective and any attempt to be photorealistic. As he increasingly widened his view over the urbanscape, he let the inevitable optical distortions become apparent and the freeway ramps torqued and bulged into a neo-surreal aspect. Given the horizontal scope of a city almost too large to comprehend, it was a visual strategy anyone living here could appreciate.

The culmination of Doolin's career came with a public art commission for the Los Angeles Metropolitan Transit Authority's headquarters behind Union Station, the railroad passenger terminal in downtown. It is in these four paintings done during 1995–1996 that everything came together. Once again Doolin charted a helicopter to survey the views from around the station, and he hired research assistants to scope out the history of the city's growth. Not only would his aerial views combine the views of the fish and bird, and rotate around the points of the compass, but they would also move through time. The paintings would trace the history of transportation in L.A. from *"Circa 1870"* through *1910*, *1960*, and project forward in *After 2000*—but also progress from early morning to early evening. His palette would shift from light pastoral greens in the first painting, which included a train passing through orchards and into town, through the increasingly murky industrialization of the early twentieth century, the smog-scrimmed freeways of 1960, and then the purples and oranges of the next century when the city has become, in his words, like a "huge, organic circuit board." Where the first painting in the series has a flock of birds traversing the deeply forced perspectival space, the final one features an airship hovering over downtown and a helicopter at the level of the aerial vantage point. You can still recognize individual buildings, and the colors aren't pushed that far beyond reality, but it is definitely the world as an artificial landscape. His painting had moved from representation of the urbanscape into using its infrastructure and setting as allegories of Manifest Destiny, environmental challenges, and a hopeful yet cautionary look at the future.

Ed Ruscha was born five years later than Doolin and in Omaha. As a kid he made a model of the part of town where he delivered papers, houses built to scale as seen from above, an expression of his early interest in architecture and aereality. When he moved to Los Angeles in 1956 he fully intended to become a commercial artist, and to that purpose had enrolled in the Chouinard Art Institute, a training ground funded by Disney where young illustrators and animators were turned out for Hollywood. Ruscha graduated in 1960, just as the school was morphing into the California Institute of the Arts, or Cal Arts, now one of the best fine arts schools in the country. Ruscha had likewise undergone a transformation, and although he walked out of school with a set of skills not unlike those of Doolin, he was under the sway of Jasper Johns and his paintings of readymade objects such as targets and maps, and the letters and numbers found on building blocks and stenciled on crates.

During the 1960s Ruscha assembled what have since become touchstones of American urban typologies, photographic catalogs of everything from gas stations and apartment buildings to swimming pools. The earliest one may have been his "rooftops," a series photographed one afternoon from the top of a building on Beverly Boulevard and looking down onto the streets below. Taken from a few feet back of the building's edge, they have a clandestine, covert feel, a hint of voyeurism that is almost always implicit in the aerial view that is made from less than a thousand feet above the ground. Ruscha also started making little chapbooks of his typologies, the most famous of which is *Twentysix Gasoline Stations*, black-and-white photos of the stations that he and musician David Mason stopped at along a road trip from L.A. to Oklahoma City. He wanted these collections to avoid statement and style, and his photographs have a deliberate flat appearance that is considered more documentarian than artistic. Los Angeles being what it is, in 1967 he hired a helicopter pilot and a photographer to fly over the city to collect images for *Thirtyfour Parking Lots*, which he published in 1967. They flew on a Sunday in order to find empty lots at places such as the department stores along Wilshire, Dodger Stadium, and the Hollywood Bowl, the white parking stripes and oil spots creating an entirely unsuspected pattern language of their own.

Ruscha professes a lifelong fascination with aerial images, which he relates to his tabletop model of the paper route. Even in the 1960s, as he was traveling to cities such as New York, Florence, Venice, and Rothenburg, he was taking pictures out of hotel windows of the streets below. He collects aerial images of Los Angeles, and spent a not inconsiderable sum purchasing a seminal aerial photo, the Berenice Abbott picture of Manhattan skyscrapers *New York: Night View*, done in 1932 from atop the Empire

State Building. Its dense black ground is covered in a matrix of lights, a vertical grid of skyscraper windows versus the horizontal plane of L.A. He also acknowledges the influence in his street maps of Mondrian's famous painting of Manhattan, *Broadway Boogie Woogie*. It's the title that transforms what would otherwise be seen as only a formalist abstraction of colored intersecting bars into an aerial image, which encourages the viewer imagining cars moving along the lines. Ruscha was using generalized elevated views as a background for phrases painted over them in his "City Lights" series. The streets appeared at a diagonal, which provided a sense of depth and motion in direct contrast to the anonymity of the backgrounds and generic language of the titles *Talk Radio* (1987) and *City Boy* (1988).

Ruscha went one further in the late 1990s when he started his street maps known as the "Metro Plots," paintings with speckled spray painted backgrounds, tiny flecks of black paint embedded in monochromatic surfaces that evoke automotive trunk paint or asphalt. They're crossed with thin lines that intersect, street names appearing in faint black block letters paralleling the lines. *Sunset-PCH* (1998), *Hollywood to Pico* (1998), *Crescent Heights and Sunset* (2001), and other paintings in the series employ the same axial perspective as the nighttime "City Lights."

The culture is awash in aerial images of Los Angeles made for any number of purposes, not least of which are the live helicopter telecasts of car chases. Nighttime aerial traverses over downtown have become a stereotypical opening for movies, a trope broadcast around the world. Such establishing sequences are meant to set the scene and mood. Ed Ruscha and Peter Alexander have painted the illuminated city grid from above for symbolic ends. By painting the grid at night and abstracting it down to the level of an optical phenomenon, they allude to, yet deter us from reading the images as map, thus creating a more open field of metaphorical potential.

Paradoxically, even though the photographs in David Maisel's aerial view of Los Angeles in his *Oblivion* series, which I discuss below, were made during daylight, they also fall within this tradition of nighttime aerial views of the city because we conflate day and night within them. In so doing, we feel as if we are penetrating the darkness and conducting surveillance, even though we are unable to read the images as maps. Maisel started out wanting to be an architect, and it's not coincidental that artists who picture what is below them often start out studying architecture. Common to both is an aerial imperative, an ineluctable desire to construct pattern in the landscape. Leonardo da Vinci, for example, was a renowned military architect before he painted his extraordinary aerial landscapes in 1502 from viewpoints imagined high above Italy. He was a genius of mechanical drawing, his axonometric views of buildings and towns were groundbreaking, and he drew floor plans for his paintings previous to executing them.

Maisel displayed a penchant for mechanical drawing in public schools, went to Princeton to study architecture, left to work at the international firm of Hardy Holtzman Pfieffer, then returned to school to earn a degree in art history. While there he also studied with photographers Edward Ranney and Emmet Gowin. Maisel's photographs of vines reveal his early fascination with pattern in the landscape, a predilection encouraged at Princeton. Working with Ranney increased his appreciation for the built form in the landscape, the older artist being one of the world's foremost photographers of Mayan and Incan monuments, such as Machu Picchu. When Maisel flew over the ashen summit of Mount St. Helens in 1983 with Gowin, it wasn't so much the crater of the volcano that captured Maisel's attention, as it was the effects of logging seen from the air, patterns of land use otherwise invisible.

When he returned to the East Coast from the flights over the newly ruined volcano, Maisel took with him a keen desire to photograph things not seen by other people, a difficult proposition from the ground. Maisel looked for relevant sites he could readily get to from New Jersey. He researched open pit mines through topographical maps and industry publications, hired a plane and pilot, and took off over the quarries and coal mines of western Pennsylvania. He flew throughout the mid-1980s to take black-and-white photographs of mines in Arizona and New Mexico, clear-cuts of timber in Maine, and then, with the help of a National Endowment for the Arts fellowship in 1990, returned for more work over the West, where he began to work in color. In 1993 he moved to the coast of Northern California to pursue photography full-time. Architecture was for Maisel a discipline that tended to isolate him from the world, while photography placed him more firmly within it, the latter a stance he could nurture more easily while closer to the subject matter that interested him.

Maisel considers his *Black Maps* project to have begun with the photographs of mines and clear-cuts that he made while still at Princeton, but widespread public awareness of this multi-chaptered epic of environmental devastation didn't occur until he began working over the Owens Dry Lake in 2001 and the Great Salt Lake in 2003. His revelation of the stunning and toxic hues of both sites have become iconic views, despite the fact that these disturbingly vivid photographs deliberately eschew reference and narrative. They can be abstract in the extreme and run exactly counter to the intent of most aerial photography, which is to specify and identify pattern and activity in the landscape. To that end, the photographs from *Terminal Mirage*, work done over the shorelines and chemical works of the Great Salt Lake, and *The Lake Project* over Owens, bear no titles, only numbers. Mondrian would have been sympathetic.

Meanders of poison, the rectilinear walls of evaporating ponds, tidal strands of blood-red halobacteria and pools of viridescent algae— Maisel's photos are surveillance of a depredation otherwise unavailable to us save from the air, but also nonrepresentational, a tension that forces us to look and look again. A secondary opposition arises in the photographs, as the images deal with the industrial sublime. They provide us with images of a beauty we are horrified to admit that we have created. That is, the images of environmental problems attract us even as the subject matter disgusts us. Maisel now eschews specific geographical titles because, in fact, people used to spend more time in front of his work discussing pollution than appreciating the photographs as artworks.

It is true that, if you want to understand politics, you follow the money. It is also true that water flows uphill toward money, as noted by Marc Reisner in his polemic on water politics, *Cadillac Desert*. If you want to understand the cities of the arid West, therefore, you follow the water. The Owens River, which collects the runoff from 120 miles of the eastern Sierra Nevada, once flowed freely southward to the base of the mountains north of Los Angeles. That watercourse was blocked millennia ago by a lava flow, and the water backed up into what was a newly landlocked basin,

Owens Lake. William Mulholland made note of that fact as he was pondering how to capture a new source of water for the city. The Los Angeles Aqueduct, completed under his direction in 1913, closely followed the prehistoric watercourse, and by 1928 the lake was completely drained, becoming the most prolific single-point source of particulate pollution in North America.

Maisel calls his *Oblivion* chapter a coda to *The Lake Project*, the deliberate desiccation of the Owens Dry Lake a corollary in reverse to the once arid, but now well-watered, L.A. Basin. He literally followed the water to see how the ruined environment of Owens Dry Lake related to Los Angeles, a journey made about the same time as Mike Light was progressing from flying over the Mojave Desert in small airplanes to renting a helicopter at the Van Nuys Airport to follow the river. What Maisel perceived when he reached L.A. was a landscape ruled over by paranoia. Hovering over the city after 9/11 was not as easy a proposition as it had been over the playa. It's fine for the police and newspeople to do so, but for a photographer it's a situation made much more tentative by new homeland security rules and regulations. He had learned from Gowin while working over the "red zone" of Mount St. Helens how to be persistent with authorities, however, and he was granted permission to traverse the Los Angeles Basin from ten thousand to twelve thousand feet, which gave him sufficient altitude to make both vertical and oblique photographs of the territory below.

When looking at the black-and-white negatives afterward, Maisel realized that he preferred the way they looked without being printed in positive. Printing in reverse, thus making a print of how the original negative is seen, provided him with an object that appeared to be less mediated within the photographic process (when, in fact and ironically, it's actually a more manipulated process). The resulting prints are as ashen as the slopes of the ruined volcano; spooky, post-apocalyptic, complete objects in themselves without being reversed back into a normative view. And they hold your attention longer as your mind attempts to make sense out of them. We expect aerial photographs to function as maps, to bear a one-to-one correspondence with reality, and the dissonance between our expectations and what we see presents us with a puzzle we are compelled to address. You immediately recognize the forms and the organizational logic of a city—the geometry of the freeways, for example, or the cul-de-sacs of a housing development—but the circumstances are strange. How can those white bands next to the overpasses be snow in Los Angeles? Oh, wait . . . are those are the foliate tops of trees that would normally appear dark? Or oddly translucent shadows? Stepping back from the image, the vertical views taken from so high up reduce the city to the recursive density of an elaborate computer chip. In negative, the darkened rooftops recede visually below the lighter walls, hollowing out every structure into empty building blocks. If the warm light of Doolin's organic circuit board offers a possibility for redemption, Maisel's brooding reversal produces a completely mechanized version without much hope.

The oblique views are even more distressing, as what is abstract is put abruptly into the context of a landscape. The circuitry of the city is suddenly linked to a shoreline, a range of hills. The skyscrapers of downtown are now axonometric figures in a virtual reality, as if displayed in a computer-aided design (CAD) program, and the course of the Los Angeles River is indistinguishable from those of the freeways. The oblique view labeled *Oblivion #1364* (Plate 5) lays out a city as elaborate as that of a Persian carpet, yet it is as white as if it had been incinerated. What presumably is a blown-out, overexposed sky above the hills in the background is instead a black void that glowers over the city. It's how an x-ray looks, how we imagine the military sees the monochromatic world when surveilling it at night. It's as if we are seeing what the artist refers to as a "shadowland," a place previously unobserved that coexists with its daylit version. It's the space of what former UCLA architecture professor Anthony Vidler called the "paranoiac space of modernism."

The oldest surviving permanent photograph, made by the French inventor Joseph Nicéphore Niépce in 1826, is a picture he took out of an upper-story window of his country house in Burgundy. The process he used was heliography, which required hours of sunlight falling on a pewter plate to fix the image in particles of bitumen. To use a French word coined at roughly the same time, it was a long "surveillance" over the rooftops, perhaps eight or more hours, and the resulting photograph was still underexposed. The low level of detail in the image leads you to feel as if you are viewing something inexplicably private, as if you were looking down into a space not meant to be spied upon.

Daguerre, who was briefly in partnership with Niépce, subsequently discovered that silver iodide was much more sensitive to light than bitumen, and in August 1939 arranged to have his friend, the scientist and former military spy François Arago, announce his invention as the Daguerreotype. Within months Arago was proposing the use of aerial photographs as a visual aid in the drawing of topographical maps, an increasingly important tool for both land development and military campaigns. It is hopelessly ironic that suburbanization has been characterized as "making war upon the land."

The term "shadowland" that Maisel uses when discussing the *Oblivion* photographs is appropriate. When you cast a shadow on a fact, you create doubt. When you shadow someone, you follow them invisibly. Shadowland is what the military calls those blacked-out areas where they wish to operate unseen, whether they are testing an experimental aircraft or interrogating people beyond lawful means. It is a land of spies and spooks, a place where ghosts live, and what Los Angeles looks like in *Oblivion*. The city is almost recognizable in Maisel's negative prints and yet not quite, as if we are seeing both more of what we know and less. It is a disquieting effect, not least because, according to Jean Piaget, a child's first conception of shadows is that they are fragments of night, which is a black cloud. The word "paranoia" crops up often in reaction to the *Oblivion* series, as if we see a shadow but can't find what casts it, another confusion suffered by children as they try to understand the nature of shadows. We look at Maisel's photographs and instinctively wonder what is casting the shadow. This exposes the essential, existential fear central to modernity: the anxiety of estrangement from the world.

The aerial work by Maisel and his colleagues gains importance as the elevated photographic surveillance of the Earth, initiated by Niépce, becomes ubiquitous. Beyond the ground being photographed from helicopters and UAVs, we can now image most of the Earth down to street level from our home computers, and the aerial, instead of remaining exceptional, has become casual. Eventually we will have seamless coverage of a fully instrumented planet, and all that will save our aerial imagination may be the work of artists. Maisel, by optically deranging his subjects, from dry lakes to watered cities, makes us stop to reconsider both the terrain and the territory. By subverting a cartographic tradition, indeed, by turning it inside out, he constructs representations that are based on the same terrain as maps of the area, yet they force us to reconsider what we think of as familiar territory, the place we have made where we live.

Maisel's *Oblivion 1364* and others in the series were taken from ten thousand feet up in both the vertical and oblique, and I like this image in particular because it allows me to compare it with that compound mural image from Çatalhöyük, both having a grid below us and the mountains in the background. But I have always read the mural to have been about the clarification of space, versus making a mystery of it. Maisel photographed from so high up for several reasons, one of which is that it's impossible to otherwise capture a built environment as large as the L.A. Basin but also because it extracts us from the folds of the landscape. James Turrell, who was a pilot before becoming an earthworks artist, notes that when you're around one thousand feet above the ground you are still in the fold of the landscape. That has everything to do with human stereographic vision and how we perceive distance. When you get above 1,500 to 2,000 feet your eyes are no longer far enough apart for our stereoscopic vision to gauge distance.

Maisel, in working so high up, has removed our ability to instinctively gauge distance and scale. Further, there are no people, no cars, no signs of life. And then he printed this series in reverse, so we have a negative image that further distances us from the environment in which we live. We are compelled to look at these images because they mystify us, and if there's anything that transfixes human cognition, it's a landscape we have trouble understanding. What Mike Light and David Maisel have done with L.A. are photographic bookends in a way, the former shooting into the light, the latter turning the light dark, both of them pushing the boundaries of our perceptual apparatus in order to make us look more closely at where we live.

# FLYING WEST
# TO EAST

*By example, we are airborne. How do you detect move-ment in a flying figure when evidence of the earth is absent? Up, down, right, left? Do you take your direction from the glance of an eye, from an eye-line, by the position of the figure in the drawing space, flying into and out of the empty page-surface? How high is a flying figure if you cannot see the ground? The need to ask such questions is because all our distances were first measured by our feet walking on the ground.*

—**Peter Greenaway**

I'm flying across the country from Los Angeles to Atlanta, then on my way to Williamstown in western Massachusetts where I'll spend the summer pondering aereality and art. The 17.2 inches of urethane foam seat I'm allowed is one of 180 seats on this Boeing 737, a long-distance narrow-bodied jetliner operated by Delta out of its Atlanta hub. The 737 is the most popular commercial aircraft ever built with more than five thousand of them in service and another six thousand or so on order. According to Boeing, one of them lands somewhere in the world every five seconds.

We're cruising at an altitude of thirty-six thousand feet and with the jet stream on our tail, as it almost always is in the northern hemisphere when you're headed west to east; our airspeed is around 550, but with a 100 mph boost from the wind, nearly 650 mph across the ground. My flight is—with a tip of the hat to the National Bureau of Transportation Statistics—one of 10.5 million domestic and international flights over the U.S. this year. I'm one of the 744.4 million passengers filling a seat on what are mostly full domestic aircraft these days. Worldwide? According to the annual report from the Airports Council International, airports handled 4.4 *billion* passengers in 2006. Presumably that is counting every single leg of each trip separately, as the International Air Transport Association, the other source of statistics on air travel, calculates that just over two billion people flew in that year. Either way, the figure is staggering. And it's growing, mostly due to flight traffic inside India and China.

When I talk about the aerial becoming a normative view of the world with the growth of commercial air travel, I have rafts of statistics to back me up. Global tourism revenues in 2006 accounted for around $2 trillion—that's $2,000 billion. Our species travels incessantly when we can afford it, and we increasingly choose high-speed conveyances through the atmosphere. I made my first flight in 1959, a ten-year-old being reluctantly shuttled off by my recently divorced mother in Reno to my father in Los Angeles. The plane was a two-prop DC-7, the last propeller-driven plane Douglas would make before the advent of the DC-8 jet, and it had been built in the plant I'd flown over with Mike Light and Denis Cosgrove just a few weeks ago.

I changed planes somewhere in California, probably San Francisco, and the second flight was at night. We flew through a thunderstorm with lightning bolts crashing around us and the plane shaking with audible thunder. I thought it by far the most exciting thing I'd ever done, and am embarrassed by the amount of carbon I've helped dump into the atmosphere ever since, having long ago surpassed a million miles in the air.

Today I'm jaded enough by the view from a commercial flight over the United States that I sit in an aisle for at least the illusion of more room—but can't help leaning over to look out the window once my neighbors get up to wander about. Fields. Lots and lots of square fields ruled off in the great American grid promulgated by Thomas Jefferson, a design descended from Çatalhöyük and the ancient cities of Mesopotamia. I remember the first time I saw this midwestern grid from the air, as a teenager in the 1960s flying in a Boeing 707 jet, and how astonished I was that humans could do something so perfectly regular to the world. It had never occurred to me before that we could shape the surface of the earth at such a scale. That view was the beginning of my fascination with the view from above.

So, as I said, lots of statistics back up my assertion that commercial air travel, in particular its growth post–World War II, is what made our view of the world from above so popular. When the DC-2 was introduced in 1934 and then the DC-3 two years later, they helped commercial air travel grow from 474,000 passengers in the U.S. in 1932 to almost 1.2 million in 1938. Most of them were business travelers or upper-middle class tourists. During World War II the aircraft industry employed more than a million people who built three hundred thousand planes for the military. That momentum was transferred in large part to the commercial sector when the war ended. The figures seem to coincide nicely with my personal experiences, and that is definitely a problem, since my conclusion is, as I am reluctantly conceding, full of holes. The aerial was with us long before airplanes, not just as a cultural predilection for bird's-eye views, but as a natural one encoded in our genes.

James Blaut was a geographer at Clark University in Worchester, Massachusetts, when he and Meca Sorrentini, a developmental psychologist (and his wife) had the brilliant idea in 1967 that they could improve the teaching of geography by helping kindergarten children learn how to read aerial photographs. Much to their surprise their first experiment was a dismal failure. The children were already adept at the task. Blaut started a small research group to study the phenomenon of what he came to call "natural mapping," and flew children, his colleagues, and graduate students around in the small plane he kept at the local airport for research purposes. Blaut ended up at the University of Chicago working on a multi-volume critique of Eurocentrism in history, but he continued to research spatial issues and in 1990 published one of the few papers to address how we imagine the aerial.

Much research is done about how we see from above looking down, impelled not least by the military seeking ever better ways for unmanned aerial vehicles (UAVs) to track down perceived terrorists, surveil them, and increasingly launch attacks against them. NASA has its own contractors— many of who also do contract work for the military—working on sets of algorithms to help UAVs on Mars conduct reconnaissance for features of scientific interest. Both soldiers and scientists spend large amounts of time and money parsing how everything from bees to hawks navigate, run search patterns, and discern pollen or prey. Much of the research has to do with boundary contrasts, pattern discernment, and some of the same principles that govern how humans look at landscapes, photos of landscapes, and maps. If that seems a bit circular, it's meant to.

A short history of vision is in order. Multicelled organisms developed sensitivity to light—rising in the oceans to follow it while the sun was up, falling back down at night—somewhere between 1.5 billion to 600 million years ago. Roughly 600 to 220 million years ago our neural systems specialized into what we can call brains and eyes. The human visual system that we have today was fine-tuned prior to and during the Pleistocene era, which lasted from 2 million years ago until the end of the last Ice Age about 11,500 years ago. By that time up to half of our cerebral cortex was devoted to processing visual information, with 80 percent of everything we learn coming in through our eyes, an unusually high percentage for a mammal.

We're so adept at receiving light that we can, given ideal conditions, see a single candle burning twenty miles away. And we're so good at recognizing figures from ground, and seeing patterns in motion, that we can find a black cat in a virtually black room. Our eyes are assaulted by one hundred million bits of information per second, enough to cook our brain if it tried to handle all of it, so a severe triage takes place. To start with, the human retina will only transmit up to ten million bits per second, and everything we see is shuffled immediately into two dozen geometrical forms. The shapes we perceive most easily and quickly are closed ones, and the more basic and symmetrical the closure, the faster we perceive it; circles are easiest, then squares and rectangles, then polyhedra such as triangles and pentagons, and finally open angles and arcs. The more complicated the shape, the longer we have to look at it, a hierarchy based on survival needs.

The brain assembles these two dozen shapes, which are formed by the contrast between light and dark, into patterns it constantly compares with those it has stored previously, mostly in our long-term memory, but some encoded in our genes. Templates are shapes we learned to pay attention to through experience. For example, if we see the shapes that aggregate into two eyes above a mouth, we pay very close attention as such a configuration often means a predator, or another primate, hence a potential competitor. Boundaries and shapes combine to form patterns, and our survival as a mammal not at the top of the food chain on the African savanna depended on our being able to discern a striped quadruped hiding in the grasses. How we see adapted to survival in that temperate environment, a process leaving us with specific protocols and hierarchies in how we process the world around us.

The mind continually scours the world for contours, and it seeks closed and stable visual elements against which to measure distance from our bodies, as well as rate of motion. The mind demands and receives three kinds of visual information: What an object is; where it is; and, what it is doing. The first it figures out through the recognition of contours. The second it constructs through a variety of perspective perceptions, including those derived from our stereoscopic vision and the elevation of our eyes above the ground, and through the breaking of contours by objects as they sit or pass in front of each other. Motion is likewise constructed from its own set of visual rules. The longer you look at something, the more significance it gains because you are cumulatively reinforcing pathways through your neurons. If you're accustomed to a specific landscape, you look for the features you've engraved most deeply into your brain.

Processing visual information, because it is so prodigious, takes up a lot of energy. Our visual systems have evolved to allow us to locate food without being eaten as efficiently as possible, and in essence that is what the military and NASA are both engaged in: trying to conserve weight and fuel while aloft in order to find whatever it is they're looking for. In the case of NASA practicing Mars, one of the programs they've been testing involves how hawks scan for prey on the ground below. It turns out that the raptors spend most of their time looking at the boundaries between areas that offer lots of cover, say the edge of a forest, and open ground. Small mammals hide in the forest but have to venture into open ground in order to forage. So the birds don't waste fuel running squared-off search patterns over fields, but instead pay attention to the edges where the caloric pay off is higher. You can imagine how that might apply to drones surveilling troop movements, or to a UAV on Mars. If, as a geologist, you know that evidence of water is most often displayed where the terrain goes from horizontal to vertical, you template the visual systems of the craft to pay more attention to such features than flat ground.

So vision is a matter of nature and nurture, a physiology that over time produces hardwired templates augmented through our individual and collective experience. How does the view from above fit into such a scheme that's based on bipeds strolling about the savanna? Back to the geographer in Massachusetts.

Here's what James Blaut and his colleagues teased out of their subjects, aged three to six, from three very different cultures in the U.S., Puerto Rico, and St. Vincent, a volcanic island in the Caribbean. First, seeing the world as if from above and interpreting how we relate to it is an inherent ability in children, one that is present in all those cultures, even when the kids hadn't ever seen anything of the world from above, or even been exposed to so much as a television or aerial photo in a magazine. By age four, the kids could "navigate realistically" with a toy car on an aerial photo set out on the floor, and by age six could identify features on aerial photos with about the same level of ability as an older child. The skill was already developed, a finding confirmed time and again by different groups replicating their studies in different parts of the world.

What Blaut concluded in response to these results was that children took their ability to physically manipulate and thus learn physical objects through touch—their microenvironment—and scaled it up to the macroenvironment around them. It's how they made place, at first understanding their immediate surroundings, and then increasingly larger parts of the world around them. They took their innate sensory skills and applied them to objects, then places. Part of what they did was first rotate objects in relationship to themselves, and then rotate the larger environment, which put their mind's eye, their imagination, above the ground.

We subject the microenvironment, anything smaller than ourselves, to all of our senses. We touch it and push against it, and it resists to one degree or another, or even pushes back. We smell it, listen to it, and taste it even as adults. Watch geologists in the field. They can't resist putting rocks in their mouths and for a good reason; there's information to be gained about their chemical composition and the physical processes that have shaped them, such as water. But the way we apprehend the macroenvironment, everything larger than ourselves, is increasingly subjected to fewer senses the larger and farther away it is from us. You can't taste the horizon. You might be able to smell or hear something from some distance, but it's mostly a matter of using our eyes. And the smaller or shorter you are, the more important that is. Children learn to cast their eyes over distance very early on.

So, unlike an object that you can hold—turn around and roll over in your hands, thus learn from all sides—the world at a distance must be imagined in order to be learned. Small children learn to rotate the world in their minds, to see it as if from above, to assemble an aerial picture of it. They do it as a matter of play, the activity that involves more of the brain simultaneously than any other activity, including sex. That's why the aerial photos made sense to them, and why they could drive toy cars along the roads. And that's why adults in cultures without photography or printed maps can relate to aerial photographs. The view from above fits our way of how we conceive of the world. Which is, of course, why we make aerial photos. More circularity, I know.

Blaut went farther, finding that preschool children had a suite of related abilities. They could rotate the macro to an overhead perspective, reduce the world to a toy or model scale, and then read simple icons representing the landscape. In short, they had protomapping skills. They couldn't make or read maps where you needed a legend to decipher what a map symbol was, or understand abstractions such as contour lines, but they could read and use simple maps before they could read text. And that meant they were well along the way to forming a system of point-to-point navigation along with increasing their mobility skills. Because these skills were developed by a certain age regardless of where they lived, Blaut proposed that perhaps humans weren't born with only a "language acquisition device," as famously proposed by Noam Chomsky in 1965, but also an analogous "mapping acquisition device." The fact that such a name provides the acronym MAD describes in a word how some cognitive scientists reacted to both Chomsky's and Blaut's proposals. Chomsky, who was fond of pointing out that children point before they speak, supported Blaut's hypothesis.

Blaut noted that cognitive mapping and place-making behavior must have evolved prior to even the manipulation of objects, which developed relatively late in primates. This allowed him to argue that humans have practiced map-making since at least the Upper Paleolithic, and he cited the Çatalhöyük mural as evidence of an already advanced and socialized skill. Blaut was careful to point out that he was not suggesting that ontogeny recapitulates phylogeny—he wasn't about to fall into the trap of suggesting that an individual develops in parallel with how a species (or a society) evolves. But he did note that map behaviors are exhibited in both playing with toys and storytelling amongst children. Because playing with toys—using small objects to represent larger ones in the real world—and storytelling are universal, he thought that maps were made not just by nomadic cultures, but also by sedentary ones, and that their purposes varied from seasonal migrations, as with the Inuit, to the maintenance of social structures, as with Australian Aborigines.

The geographer and his wife believed that schools should build aerial photographs and maps into their floors in order to promote their innate mapping proclivities, and that field trips should include airplane rides so that they could connect their protomapping skills to the landscape. When I read his proposals, I remembered that the photographer David Maisel talked about how his playing with blocks as a child led him to want to become an architect, and how that kind of elevated view of the world transferred into making aerial art.

You can see why I use the Çatalhöyük map as a fundamental image in books and lectures, as it demonstrates at a relatively early stage in the development of human culture how transforming land into landscape, or terrain into territory, is a process first engaged through our natural neurophysiology and then taken over by cultural means once we hit the limits of our senses. And so, of course, Blaut's work appeals strongly to me. I consider the God's Eye view as an evolved adaptive strategy favoring the survival of those able to practice it, insofar as it allows us to imagine conditions around us, in particular to predict what is distant spatially and therefore potentially ahead of us in time. That's why we climb a tree or mountain to ascertain where we are and what path we might take. I would add that, because we're hominids that climbed down out of the trees sometime between five and six million years ago, arboreal movement and thus aerial perception are still very much with us. Swinging from branches and jumping through the air from tree to tree in an effort to stay above the ground and thus avoid predators is why we have opposable thumbs and stereoscopic vision. That's also been cited as a possible reason why humans in all cultures across history have dreams of flying.

It's tempting to turn this into a parable by saying an arboreal existence was our prelapsarian state, that coming down into the savanna on clumsy feet was the original fall, and that the God's Eye view is, among other things, an attempt to reclaim that state of aerial grace. That would be more metaphor than fact—a metaphor being a specific kind of analogy, an unexpected comparison—and it allows me to segue from thinking about how the aerial view is a mental ability that we use to demystify the world, and to argue that we also use aerial views to remystify the world by seeking out "news of difference."

When I started researching this topic, I thought that the aerial images made during World War II, followed by the ubiquity of commercial air travel in the 1950s and '60s, had made the aerial view of earth a normative one. Google Earth, which appeared a year later, pushed that viewpoint, in both senses of the word, beyond anything I had imagined. But as with the histories of all landscape representations—from maps to paintings to ground-based photographs—the uses for the images undergo change. This was related to what I'd talked about with Weston Naef during our lunch in 2006 about my work. As I related earlier, the Getty had just bought and was getting ready to exhibit Garnett's Lakewood photographs, and I wondered how Weston felt about aerial work in general. His response was that aerial photography was fun, and that he loved to scroll through Google Earth. This was the basis for his difficulty with it.

Aerial photographs are fun and that's a problem. You can't help looking at them. They're like the snapshots of dead people that you find in a flea market. I never met a snapshot I didn't like. But art isn't fun. It's something else. Fun is entertainment. Diana Arbus and Gary Winogrand had to confront the snapshot and stylize it. That's the challenge. And to get there first. The William Garnett pictures—he got there first. We bought them also because they are relevant to our territory.

Google Earth overtook aerial photography. Now you can call up almost any longitude and latitude and get a picture of it. I love scrolling through Google Earth until I come across an area that's been blacked out by a government. That's the only way you come across the remaining secret places on Earth.

But, of course, looking for secret places is a game, Naef was saying. He ruminated for a minute and then added: "It will have to be the next generation of kids who grow up with Google Earth who will meet that challenge of doing something that's not just fun, but art."

Well, I agree that art has to be more than simply entertainment, although good art can certainly be fun. But I would submit that artists have been mak-ing aerial views for a long time, images that are beyond entertainment and are an important, if vastly understudied, part of the formal conversation of the larger art world. Aerial artists in general, but most particularly photogra-phers, are still laboring under the curatorial misapprehension that they're doing map work, the traditional complaint leveled against landscape paint-ing by the gatekeepers of art academies in the eighteenth and subsequent centuries. The more ambitious landscape paintings of the nineteenth century somewhat escaped that ghettoization, thanks in part to J. M. W. Turner. Turner managed to remain friends with Sir Joshua Reynolds, head of the Royal Academy in London and one of the more conservative gatekeepers, even as he turned landscape work on its head; perhaps that's exactly why he retained the older man's respect. One of the reasons I'll be flying over the Hudson River Valley is to examine that shift and the role of the aerial in it.

Likewise, landscape photography slowly gained respect during the twen-tieth century, but most aerial photographers, even as they make the best-selling and most widespread landscape depictions in the world, are considered more as makers of mapworks than of art, as Sir Joshua Reynolds put it. Personally, I found Google Earth all the more reason for aerial photog-raphy to be studied, written about, and collected precisely because it now made the God's Eye vantage point interactive, though only within a very lim-ited frame. And the best of the aerial photographers, such as Mike Light and David Maisel, were busting the frame all to hell and gone.

Google Earth is a virtual global imaging program first developed by Keyhole, Inc, a company named after the original military surveillance satellites first launched in 1959. The company was acquired by Google in 2004, its product renamed and released in 2005. When you launch the program on your desk-top it begins with a view of the entire Earth as if from space. You can start as far away as thirty-nine thousand miles, and then using a combination of GPS data with satellite and aerial images, zoom down to a resolution that in many cases equals fifteen meters, or about forty-nine feet, per pixel. Over some cities the resolution is as high as fifteen centimeters, or six inches per pixel. Looking at my house in Burbank, for example, I can count individual paving tiles that are twelve inches each across. People walking down the street stay focused until about 650 above the ground. The resolution is good enough to see how your lawn is doing, but not enough to identify individuals. You can rotate and tilt the images just as you would if you were one of Blaut's five-year-olds mentally manipulating the spaces inside your house.

Google Earth does have inherent limitations, some of which are very deliberate, some not. Apart from the fact that it uses images at least a couple years old in order to protect privacy, and it is required by certain security concerns to leave some areas fuzzed out, the technology will make the terrain surrounding singular tall structures, such as the Eiffel Tower in Paris, pull up into an apparent hill when none exists. But the main limitation is that the service can only use a general perspective as the basis for the imagery—that is, vertical views facing, in essence, the center of the earth under the selected location of interest. You can make it seem as if you're flying over three-dimensional terrain, but it's all based on mathematical projections done from those verticals, along with elevation information. It's very cool, but the views are definitely map-like, which is to say static and without any of the resolution or framing that the human eye and mind provide.

So here I am, looking out over Middle America's Great Flat from thirty-six thousand feet. Green and brown squares and rectangles are intersected at right angles by property lines, which run parallel to roads that jog every so often to account for the fact that they were laid out along surveyed map lines, which work fine on a two-dimensional surface, but every few miles on a spherical surface have to be adjusted to continue along straight and true to the actual shape of the globe. It's pretty, but when I imagine myself on the ground, pretty boring. I'm not a fan of flat terrain and the mental territory that, stereotypically, popular culture would have us believe accompanies it. And then I'm rescued from the clichés as I think of the photographer Terry Evans and the work she's been doing over the midwestern prairie since the early 1970s. She's someone else I should discuss with Weston, an example of how aerial photography doesn't just simplify the world into cartographic patterns, but re-complicates it for us, re-mystifies us.

Our historical aerial conception of the Midwest stems from a variety of sources: bird's-eye view artists, nineteenth-century atlases, early twentieth-century barnstormers, commercial air traffic, and maps. But the government had a surprising hand in it, as art historian Jason Weems has outlined. Aerial photography, by virtue of its ability to reveal the larger context of land use practices, was adopted by the government during the 1930s as an important means of combating the effects of the Dust Bowl. The Farm Security Administration (FSA) was created in 1935 to address rural poverty, which was particularly acute during the Depression. Its photographic division, under the direction of Roy Stryker, was charged with the task of familiarizing Americans with the problem, a political necessity when applying public tax funds to the issue. Stryker, although not a photographer himself, helped shape mid–twentieth century photography by hiring artists such as Dorothea Lange, Walker Evans, and Gordon Parks to document the plight of agricultural workers

The Midwest had already been perceived from the air through the popular bird's-eye views of the previous century, and barnstorming flights given to locals by ex-military pilots working the county fair circuit. Stryker, who knew Bourke-White, had photographers in the air as early as 1936 to take pictures that would give context to the people and towns. At the same time, the Agricultural Adjustment Administration (AAA, a precursor to the U.S. Department of Agriculture) began to conduct aerial photo surveys over farmlands in the Dust Bowl to check farmers' compliance with production and then soil conservation controls. According to geographer Mark Monmonier, the farmers mostly accepted what some could have deemed surveillance by a hostile agency; after all, the FSA was using methods based on those developed by the military for aerial photo interpretation. Perhaps because the monitoring was tied directly to cash subsidies for land held out of production, resistance was muted. In 1937 the AAA had thirty-six crews in the air photographing 375,000 square miles, and by 1941 they had captured more than 90 percent of the country's fields. From its initial goal of promoting compliance, the Agriculture Department's aerial photography program became a tool for conservation and land planning as well as an instrument of fair and accurate measurement. Local administrations and a widely perceived need to increase farm income fostered public acceptance of a potentially obtrusive program of overhead surveillance. Soon thereafter, World War II became the more important battle, and the military absorbed both Stryker's unit and the AAA personnel and its two photo laboratories. In a nice bit of symmetry the military then was taught aerial imaging techniques refined by both agricultural agencies over the Midwest. But by then the precedence for using aerial vantage points from which to analyze and design land use practices had been established.

Stryker, who was born on a farm in Colorado, but who majored in economics at Columbia University, understood relationships between the micro- and the macro views of landscapes. He promoted the work of Lange, Evans, and Parks (and by extension, that of his agency) through *Look* and *Life* magazines. When the FSA was disbanded in 1943, the photo unit was absorbed by the Office of War Information, and Stryker insisted the 130,000 photos of the agency be housed at the Library of Congress. In 1962 Edward Steichen curated a show from them at the Museum of Modern Art titled "The Bitter Years: 1935–1941."

Terry Evans, born in Kansas City in 1944, didn't start out working from the air, although photography was an early skill. Her parents ran a professional wedding and portrait studio. Her father gave her a Kodet camera when she was four, and she still has a picture of him holding his Speed Graphic. She was attending the University of Kansas as a drawing and painting major when he lent her a Nikon to photograph the 1968 Bobby Kennedy rally where the young politician kicked off his presidential campaign. The event was a heady one and her pictures came out well enough to convince her to continue photographing. By 1974 she was working with the renowned photographer and curator James Enyeart, photographing farm people for a publicly-funded project about "Kansas in Transition." In 1978 she discovered the undisturbed prairie and began to survey it through her camera held at waist-level.

That early work shows her absorbed in the pattern of grasses, how a clutch of feathers where a bird has been preyed upon splays out, how the smallest of flowers makes room for itself and orders the ground around it. But she was also pointing the camera outward, the square format of her landscape photos divided neatly in half by the horizon. And she was pointing upwards and photographing patterns in the clouds. By the mid-1980s, she was flying above the prairie. Through a series of interviews and e-mails, I've been asking her about the transition. "The aerial work from waist-level started in 1978 . . . I was photographing virgin prairie in Kansas, trying to look at a complex ecosystem and show the interrelatedness of it, and looking down to show the relationships among plants. There must have been 200 species—this enormous biological complexity of the prairie. I had not done any aerial photography before, but I wanted to see if the same patterns would emerge from 500 to a 1000 feet up, and they did, of course."

Her first book of fifty-nine color and black-and-white prairie photographs, *Prairie: Images of Ground and Sky*, shows how her work from 1978 through 1985 revealed a surprising complexity in how the mind constructs that space. The tightly cropped close-ups and the aerials are sometimes difficult to distinguish from one another. The clouds become fields, the fields look like sky, and in one photograph you can rotate the view upside down and not be sure what your orientation is.

Evans was uniquely well prepared to take on what most people consider to be a highly isotropic environment, in essence a monolithic space that looks the same in all directions. Her favorite teacher at Kansas was the Yale-trained painter Peter Thomson, who taught her " how to move through the space of a drawing. I mean I learned how to draw so that the eye could move through the drawing and so that I could show many vantage points in the same draw-ing so that the eye could see several different spatial planes. I loved drawing and it was then my greatest strength. . . . I admired the California School painters like Diebenkorn and Thiebaud." Both are painters who deploy an aerial per-spective to reconfigure the landscape, and offer lessons in how we look at it.

Two people wrote introductory essays to the prairie book: Wes Jackson, the founder of The Land Institute (which studies and promotes natural sys-tems agriculture), and the legendary anthropologist and systems theorist, Gregory Bateson. While the presence of the former indicates to us how deeply Evans was (and remains) involved in environmental issues, the latter allows us to open up another complex layer of meaning related to Evans's use of the God's Eye vantage point. While Jackson was beginning to work out relationships between the ecology of plowed fields with that of the nat-ural prairies, Bateson was writing his *Steps to an Ecology of Mind* (1972) in which he asked a question obviously relevant to the work Evans was doing. He starts by quoting the famous statement made in the 1930s by the math-ematician and founder of general semantics, Alfred Korzybski: "The map is not the territory." Like his predecessor, Bateson believed that humans could know only what was transmitted through their neurology and language.

We say the map is different from the territory. But what is the territory? Operationally, somebody went out with a retina or a measuring stick and made representations that were then put on paper. What is on the paper map is a representation of what was in the retinal representation of the man who made the map; and as you push the question back, what you find is an infinite regress, an infinite series of maps. The territory never gets in at all. Always, the process of representation will filter it out so that the mental world is only maps of maps, ad infinitum.

But Bateson gave us a way to step outside and above that dilemma by examining patterns within nature, including the nature of thought, a meta-pattern. Evans's work is all about looking for patterns in both time as it is made into history, natural or otherwise, and in space as it is made into place—and sometimes that unique conflation of the two, of history and place, as we try to set back the clock. What she discovered in 1978 was not so much the prairie per se, but the gap between reality and representation, between map and territory, and how to exploit that difference.

Evans published two books in 1998, *Inhabiting the Prairie* and *Disarming the Prairie*. The first, comprised of black-and-whites taken from 1989 through 1992, catalogs a complex landscape that is never exactly as flat as the clichés would have us believe of the Midwest. It rolls and dips and tucks, and everywhere there are marks made by us, subtle and not. The meanders of creeks are juxtaposed to the contour plowing that exposes the degree of slope; the careful elucidation of property lines creates empty squares, or a dense block of trees defined by the surrounding fields. Mining operations and military bases, silos and farmhouses, country lanes and interstate free-way exchanges. And always she returns to the ground. The long sheds of corporate agriculture from a few hundred feet up contrast with two horses behind a wire fence, one white, one black. The circular bombing targets and defensive emplacements around a missile silo are balanced with the crum-pled carcass of a military jet surrounded by weeds and graffiti. Compare and contrast, look for news of difference.

When writing about maps Bateson was not decrying their usefulness. On the contrary, they were essential pieces of evidence undergirding one of his most useful dictums, that "Information is news of difference."

What is it in the territory that gets onto the map? We know the territory does not get onto the map. That is the central point about which we here are all agreed. Now, if the territory were uniform, nothing would get onto the map except its boundaries, which are the points at which it ceases to be uniform against some larger matrix. What gets onto the map, in fact, is difference, be it a difference in altitude, a difference in vegetation, a difference in population structure, difference in surface, or whatever. Differences are the things that get onto a map. [. . .] In fact, what we mean by information—the elementary unit of information—is a difference which makes a difference.

The photographs of the military presence on the inhabited prairie hint at what was to come in the second body of work made by Evans in the mid-1990s, which is all about the evidence and exercise of difference.

When the United States entered World War II, the military immediately began to construct a series of ordnance manufacturing facilities around the country. Each of them needed to be inland and away from possible attack by foreign aircraft or naval bombardment, but have close access to railways lines. The Joliet Arsenal, later named the Joliet Army Ammunition Plant, opened in 1940. Eventually more than ten thousand people worked in its 1,462 buildings spread out across 36,645 acres and enclosed by thirty-seven miles of fence. Because of safety reasons, as much as 19,000 acres of the facility was left undisturbed, the tallgrass prairie serving as a buffer zone. The arsenal was, however, only forty miles southwest of downtown Chicago in 1940, and one explosion at the plant, which killed fifty-six people, was felt as far away as sixty miles.

During the war Joliet's two plants manufactured a billion tons of TNT and loaded more than 926 million bombs, shells, mines, and other ordnance. The plants continued to produce munitions through the Vietnam War, but were finally closed in 1976. As with any installation deemed redundant, plans for its future disposition were eventually considered, although pollution of the ground was severe enough that it was declared an EPA Superfund site. Some of the less disturbed land was used for grazing and farming, but in 1993 the discussion went public over how to use the 23,000 acres that the Army deemed excess. Eventually 3,000 acres would go to an industrial park, where Wal-Mart has erected a 3.4 million-square-foot distribution center. The Department of Veterans Affairs received 982 acres for a national cemetery and 455 acres were set aside for a county landfill (a relatively common, if ironic and seldom remarked-upon conjunction of land use across the country). But the majority of the property, 19,000 acres, would be reserved for an eco-reserve known as the Midewin National Tallgrass Prairie. The reinforced concrete bunkers used to store munitions were too strong to dismantle, but the offices and assembly buildings would be torn down, the contaminated dirt hauled away or buried. Evans photographed the property before all the con- temporary changes occurred, and her work is thus both art and a valuable historical document.

*Disarming the Prairie: Creating the North American Landscape* is an even more complicated look at the situation than you might suppose. The disposi- tion of a modernist ruin—and that's what the old complexes of ammo igloos left around America look like, Wendover included—is fraught with opportu- nities for theorization by contemporary geographers. You stare down at the Midwest from a jetliner at thirty-six thousand feet and think to yourself that the reclamation of an old military base for a habitat of endangered species must contain plenty of news about differences. But then you think about the land having been shaped before the war by settlers and the farmers from whom the military purchased the land. Europeans arrived in Illinois as early as 1673, and by the mid-1800s you had not just pioneers from nearby states and the Northeast moving in, but Irish and German immigrants farmsteading and building towns on the prairie. Before that, of course, the twelve nations of the Illinewek and other native inhabitants were traversing the area on hunts, and burning the grasses to promote the growth of crops. Then there's the Paleo-Indian Period that ended about eight thousand years ago, around the time when the mural at Çatalhöyük was being made. No one yet really knows for sure how far back that goes. Twelve thousand years "before pres- ent," as an anthropologist would put it, is what the government's website for the reserve states, but that is typically conservative. Evans, because she had been working on and over the prairie for years—and both prepositions are critical—managed to pay homage to many of the region's timelines.

The cover image of the book is iconic (Plate 6). In the foreground neatly rolled hay bales rest on the rows of a harvested field. Across a road and in the background, rows of tidy bunkers are set at roadside. The scene is peace- ful, shadows dappling alike both properties. Rail cars sit on a siding next to a closed bunker over which the prairie has long grown, the difference between insulation, camouflage, and abandonment only a matter of intent. Other photographs juxtapose a prehistoric Plenemuk burial mound with a nineteenth-century cemetery, both almost inscrutable ruins; or, tall barbed wire enclosures parallel a stone fence built by Confederate prisoners of war. It's not just the contours of land exposed by views of plowing that Evans made, but contours of time.

The photographs are emblematic of a larger reality, as well. Tony Hiss, writing in his essay that accompanies the photographs, reminds us that before the Second World War, the American military controlled approximately three million acres of land in the country; by the time the Cold War ended, that figure had ballooned to thirty million. Base closures and other means have trimmed that to around twenty million acres at present, about 1 percent of the contiguous United States, and an area the size of Austria.

More recently, Evans has been photographing on the ground and in the air over Chicago, what she calls "the urban prairie." The contrast could not be greater. Over Joliet the traces of humans were everywhere, but as Hiss points out, there are no faces. No people. In the Chicago work, done during fifty flights in helicopters, airplanes, and hot air balloons, the farmyards abut the suburban backyards, and clearly living goes on. Yes, she captures urban decay and abandonment—the empty lots and rusting infrastructure, the shuttered malls and closed quarries—but people are walking dogs on the beach and swimming in the public pool. There's the geometry we expect in an aerial view of a cemetery, but also a visitor. Inmates lines up in the exercise yard, boats are plying rivers, and more often than not narrative is everywhere in the act of being created. Her vantage point is, like Google Earth's, just distant enough so that privacy is not invaded, but here the aerial accounts for more than pattern. It includes people in the process of creating it.

When Gregory Bateson was writing about information being news of difference, he was basing his statement on two branches of science, information theory and neurophysiology. He followed experiments into the nature of our visual systems in order to determine the various ways in which we created difference. Among them were the fact that we have two eyes, hence binocular vision, a singular view of the world created by communication between two distinct halves of the brain; and, more importantly, we move our eyes constantly, scanning in "saccades" or arcs of between two and three per second. In one experiment, scientists attached an apparatus to the heads of their subjects that matched the movement of an object in the visual field to the saccadic movement of the eye. The subjects were unable to see the object because there was no news of difference. And, most importantly, we move our entire heads as we look at the environment around us. That creates angles of incidence—the degree of difference from looking straight onto the surface of an object, what is defined as the "normal angle," versus from the side, the angle of incidence. That difference is critical to our being able to establish the position of other objects relative to our own. That 80 percent of everything we learn every day being visual, coming in through our eyes? It's all based on difference, and it is how we are able to visually and imaginatively manipulate space and our place in it.

Google Earth doesn't move your head from side to side. It looks straight down on your roof. Even when you use the feature that allows you to tilt the image and turn the view into a three-dimensional one, you are only working from images compiled from the normal angle. There is no angle of incidence, no news of difference. Terry Evans, who most often photographs obliquely in order to deliberately include two points of interest in each frame, captures difference. For the most part Google Earth seeks to get us from one place to another by simplifying our view of the world. Evans seeks to get us from one place to another by complicating it.

Don't misunderstand me. I'm not suggesting dualities, nor even a linear spectrum across which you can say one kind of imaging technology gives more information than another. I'm not interested in trying to privilege photography over mapping or painting over photography, for example, but I am saying they exist differently. A vertical aerial photograph contains immense amounts of data for our eyes to roam across, and Google Earth, by combining satellite images, aerial photos, and maps is one of the most powerful tools ever invented. It can help you navigate in a car, or compare data sets otherwise too large to comprehend. For example, you can repurpose its content by combining it in mashups with other information. You can plot atop its maps the outbreaks of Avian flu to find out what kinds of geography favor the disease. You can take real estate listings, plot them onto Google Earth to look at the streets without even having to drive down them, then add in demographics about the neighborhoods. Google Earth can be used to create difference, but it isn't art. It's trying to make the map match as closely as possible the territory. Photographs by Evans are art because they make visual correspondences we didn't expect and that are not at all the kinds of associations we make from reading a map. They are metaphors, in short, yet another way of knowing the world. They're fun to look at, but as serious as can be.

So here I am, flying over the prairie, and wondering why it is that, if we're born with the ability to rotate space mentally and create an aerial view of the world around us, and some Aboriginal people are able to read aerial photographs without having been trained to do so—why it is that Jim Bowler's students in Melbourne have to be taught the ability? To compound my confusion, if commercial air travel and increasing numbers of aerial photographs are prevalent in the culture, shouldn't we all be adept at reading such images now given us by technology? Plenty of questions remain for me to investigate.

# BACK EAST

# 8. Time Flies By

*The sky is the soul of all scenery.*
                                    —Thomas Cole

**Joe Thompson banks steeply around the ridge in the Green Mountains** of southern Vermont so I can get a good look at the sixty-five-foot-long blades spinning just below us. The fiberglass fingers are attached to eleven wind turbines that sit on white towers 130 feet tall, and the black scythes rotate through the air once about every four seconds in the breeze. Needless to say, Thompson keeps his distance. This is the Searsburg wind plant, and when it opened in 1997, it was the largest such facility in the eastern United States. The thirty-five acres of the plant produce enough power for two thousand homes.

We circle the 2,900-foot ridge, then head south and back into Massachusetts. The blades are black so they'll absorb sunshine and heat up enough to shed ice during winter, but on this June morning that's not an issue. The cockpit of the 1979 Grumman Tiger is pleasantly warm, and I can see the skyline of Albany over on the Hudson River to the west, parts of Connecticut to the far south, and Mount Keane to the north in New Hampshire. To our near south is Mount Greylock, at 3,491 feet the highest point in Massachusetts, and the reference point I look for every morning when walking around Williamstown where I'm staying.

Compared to the mountains around Wendover and Los Angeles, which are two to three times higher and reach above the treeline, Greylock is more like a forested hill. On the other hand, given the density of the forest on its flanks, it's a terrain in which hikers find it easy to become disoriented. In fact, the ground here overall has far too many trees for my taste. For one thing it's impossible to see how many people actually live in the area. To my untutored western eyes, much of the land looks like uninhabited forest, when in fact it's threaded everywhere with roads, including to the top of Greylock. It's only when you're directly overhead that you can see the roofs of houses. From an oblique angle, the country looks scarcely populated at all.

Joe, taller than six feet and surprisingly boyish-looking for a mid-forties guy who directs the largest contemporary art museum in America, is scouting for a place where he can locate a windmill. The Massachusetts Museum of Contemporary Art in North Adams, otherwise known as Mass MoCA, consists of twenty-seven old factory buildings on thirteen acres, and it has an energy problem. As we tour over the surrounding hills and valleys, I'm amazed at how much of the landscape is devoted to power generation from wind, water, coal, and nuclear means.

The area that's now North Adams was surveyed in 1739, and a fort was built at the conjunction of the Upper and Lower forks of the Hoosic River. By 1799 the town was an important textile center with waterworks powering the mills. There was a brickyard, a sawmill, and an ironworks that provided the metal for armor plates on the USS Monitor, the first ironclad ship, which saw battle in the Civil War. The brick buildings of the museum started out in 1860 as a cloth printing factory that supplied uniform material to the Union army, and at the beginning of the twentieth century the Arnold Print Works was the largest employer in town with 3,200 workers. The Depression took its toll, however, and the facilities were bought up in 1942 by the Sprague Electric Company, an electronics manufacturer that produced circuits for the military during World War II, including components for the atomic bombs that would find their way to Wendover.

During the latter half of the century Sprague continued its government work, making systems used in the launch systems for NASA's Gemini space missions. At its height in the 1960s the thirteen-acre complex covered a third of downtown North Adams and employed more than four thousand of its eighteen thousand residents. The development of the integrated circuit—known as a chip to you and me—and manufacturing competition overseas forced the plant to close in 1985.

Mass MoCA came into possession of the property in the 1990s, and opened in 1999 with nineteen galleries and one hundred thousand feet of exhibition space. It's a stunning place in which to view art, the three-story brick buildings connected by flyways, and the rows upon rows of twenty-four-pane windows providing a visual continuity to the complex. Said bricks and windows, however, leak heat faster than a perforated t-shirt, and the museum's combined gas and electric bill has climbed in the last two years from $200,000 annually to more than $750,000—and Joe figures this is just the beginning. The only place in the museum's $5 million budget that he can cut is in programs, and as any museum director knows, that can signal a death spiral. Cut programs, fewer people come, revenues drop even farther . . .

Joe believes that the museum has to participate in the supply end of energy, not just the consumption, and would love to add a single 1.5 megawatt generator to someone else's wind farm project. Here's how it would work. According to energy analyst Charles Komanoff, writing in the Massachusetts-based magazine *Orion*, the state has ninety-six linear miles of ridgelines where winds average 14 miles per hour or better, which is what you need to generate power at a profit (although the turbines actually start to produce power at 10 or 11 mph). About a quarter of that area is low-conflict, where you're not impinging upon public viewsheds in parks, conservation areas, and so forth. Given that the local area here gets 30 to 40 percent of its days with winds good for generation, even at that lower end the museum would generate power during 8,000-some hours a year—about 4,000 megawatts—and it only needs 3,000. Thompson figures he'd break even by selling the excess 24 percent of power generated to the grid.

This is not your average museum facilities plan, but Mass MoCA has a habit of thinking outside the box, an attitude signified by the sculpture that greets you in its front courtyard: six maples trees suspended upside down from cables between telephone poles. The installation piece, Natalie Jeremijenko's *Tree Logic*, continues to grow both downwards and upwards—the trunks extending toward the ground year by year, the branches turning upward toward the light. The trunks display gravitropy, the branches phototropy, two mutually contradictory responses to the situation. It's such an unexpected sight that the first two times I walked by it, my mind automatically converted the trees rightside up, and I took the metal cans from which they grow to be transformers hooked up to transmission lines.

Aesthetics are, ironically, the central issue in whether or not Joe can build his windmill. The people of Massachusetts, specifically those with valuable second homes in scenic areas, aren't exactly enamored of power generation done within eyesight. Even such liberals as Edward Kennedy, and the environmental writer Carl Safina, for example, have been fighting the construction of a wind farm six miles offshore from Cape Cod. Joe thinks he might have a way around that: he'll commission an artist to design the windmill as a public art piece. As the museum director puts it: "You need three things. A location with wind, like a ridge, a nearby power line, and a road for access. Oh, yeah, you also need poverty. You couldn't put a windmill anywhere in sight of an expensive neighborhood."

As we tour over the Berkshires in Joe's Tiger, I look down to spy upon as many power generating schemes as I can find. It's the kind of aerial strolling Mike Light and I were doing over both the Great Salt Lake Desert and with Denis Cosgrove over Los Angeles. Contemporary geographers resurrected the nineteenth-century French word for attentive strollers of the boulevard, a *flâneur*, someone who walks slowly in order to observe with a certain level of detachment. It is said that a *flâneur* would promenade a turtle on a leash in order to maintain the correct pace down the sidewalk. The English geographer Davide Deriu has recently coined the term *planeur*, a "glider," for someone who undertakes the same activity in the air. From the casual way Joe and I are dressed, to the notebook in my lap, to the fact that his single-prop plane is flying at a stately 110 miles per hour, we qualify.

The more I look, the more power line corridors I see cut through trees. If it weren't for the incredibly thick foliage, people would have a lot to complain about. Every change in elevation has been used to impound water and generate electricity from the pull of gravity upon it. As we head south and past the property where Herman Melville used to live, we pass a reservoir on top of a mountain. "See that pond?" Joe asks. "They buy power when it's cheaper at night and use it to pump water from the river up to the reservoir. Then during the day, they let the water run downhill through turbines and sell the power they generate back to the grid at a higher price, making money." I take it that a reservoir, especially one on a peak, is a recreational plus, not a downside for the locals.

Many of the larger valleys host coal-fired power plants, all of them feeding into the grid that distributes power in the Northeast to cities such as Boston and New York. From the air I see the occasional town, but not many roads. They're under the trees. Cemeteries are prominent both because the countryside has been settled for centuries, but also because the memorial parks are mostly open space with mown lawns. We venture a bit farther south of the reservoir, and before heading back to North Adams, fly over what's left of the Yankee Rowe Nuclear Power Station, the third commercial nuclear plant built in the country and the first in New England. We're allowed to fly over the facility because it closed in 1992 after operating for more than thirty-one years. They've almost finished structural demolition. Next they'll excavate and ship off the contaminated ground, grade the site, and replant it with native vegetation, but, although it will look reclaimed, radiation will prevent people from using it for decades, if not longer.

The elevation variations and consequent vantage points offered in this part of the country also helped establish a different kind of power in the eighteenth and nineteenth centuries, a more abstract social and political kind that we refer to as Manifest Destiny, and it was expressed nowhere else more vividly than in the paintings of the Hudson River School, the practitioners of which took issue with the idea even as their landscape views were being used to promote it.


Mass MoCA is the largest employer now in North Adams, and the survival of both museum and town are largely dependent upon tourism and art, both growth industries that started here and in the Hudson River Valley in the mid–nineteenth century. The river, really a tidal estuary at sea level that stretches from New York Harbor north 153 miles to Albany, was long an important waterway for the Algonqiun-speaking Mahicans, who had observed the tidal pulse and called the river the Muh-he-kun-ne-tuk, or "river that flows both ways." The region derives much of its scenic virtue from the fact that what we call a valley is in reality the drowned canyon of a river that flowed far beneath the present-day surface. When we look at the steep thousand-foot high hillsides north of West Point, we're observing what the river cut before the sea invaded.

Henry Hudson sailed up the river in 1609 looking for a passage to the Pacific before being forced to turn back as the river became non-navigable near the site of present-day Albany. A Dutch trading post was founded nearby the next year, and the river subsequently served as a key route in support of the fur trade. Agriculture and manufacturing weren't far behind. The invention of the steamboat in 1807 made travel up the river quicker and safer, and the completion of the Erie Canal in 1825 connected the Hudson to the Great Lakes, which soon helped bring the coal from mines in Pennsylvania to New York City. The river became the most industrialized thoroughfare in the United States even as its surrounding scenery began to attract visits from the citizens of the increasingly crowded and polluted city.

That same year a young British woodblock engraver who had emigrated to Ohio in 1818 decided to become a landscape painter and moved to New York City. That summer he took a sketching trip into the Catskills, an eroded plateau southwest of Albany that featured picturesque streams and waterfalls. The European idea of the Grand Tour was already well established in New York, and Thomas Cole's trip up the Hudson followed an established scenic route. The pictures that he painted upon his return fit into this American version of a landscape tour; his romanticism presented the viewer with a wild and somewhat gothic scenery, yet the details of nature were minutely observed and accurately portrayed. The scenes were identifiably and authentically American, and brought him instant acclaim in the city.

Cole made two trips to Europe to study art and expand his reputation, and after returning from the second one in 1833, he built a house on the west side of the river in the Catskills. He had seen firsthand the work of the classical and contemporary landscape masters, notably Claude Lorrain, Constable, and Turner, and from each of them adopted techniques he could use to paint his grand moralistic series, such as the five-part *The*

*Course of Empire* (1836). Cole wasn't above idealizing the American land-scape for his allegories, but he also promulgated the idea that wilderness equaled nature equaled God, an equation with which one did not need to tamper. Along with Turner, he believed that the goal of the artist should be fidelity to nature even as he was striving for the sublime.

In the midst of painting his apocalyptic rise and fall of a mythical civilization, he took time off to paint what became the most famous American landscape picture of his time, *View from Holyoke, Northampton, Massachusetts, after a Thunderstorm*. Based on tracings he had made in Europe of a topographic sketch of the scene in a book published by Basil Hall, the view is set atop the thousand-foot hill. The view below to an oxbow, a sweeping and nearly 360° bend in the Connecticut River that almost intersects itself, afforded what the previous year he noted in his "Essay on American Scenery" was an opportunity unique to an elevated vantage point: that of enabling "the mind's eye" to "see far into futurity" when "mighty deeds shall be done in the now pathless wilderness."

The walk up Mount Holyoke with its commanding view was so popular that in 1821 a shelter was constructed on the summit for hikers; in 1851 a hotel took over the spot. Cole's painting is bifurcated between the dark wilderness on the left, signified by a broken tree and departing storm, and settlement on the right, a sunny aerial view of farms and fields. It was a prescription for what the New York journalist and editor John O'Sullivan in 1839 wrote about in an essay titled "The Great Nation of Futurity:"

> The far-reaching, the boundless future will be the era of American great-ness. In its magnificent domain of space and time, the nation of many nations is destined to manifest to mankind the excellence of divine principles; to establish on earth the noblest temple ever dedicated to the worship of the Most High—the Sacred and The True. Its floor shall be a hemisphere—its roof the firmament of the star-studded heavens, and its congregation an Union of many Republics, comprising hundreds of happy millions, calling, owning no man master, but governed by God's natural and moral law.

He could have been describing Cole's painting.

The first great master of the American landscape died unexpectedly young in 1848, but not before tutoring several other important artists. Among them was Frederic Church, who would take the elevated viewpoint as one necessary not for the proclamation of a divine right to Manifest Destiny, but for the unification of a worldview based on observation and science. Church's life and work were deeply shaped by the polymath Alexander von Humboldt, who may have been the last man alive able to grasp enough of contemporaneous knowledge to attempt writing a complete synthesis of it. His scope is reminiscent of Leonardo da Vinci's, and it's worth revisiting first the aerial work of the quintessential Renaissance Man.

Leonardo seems to have had dreams of flying his entire life, and an obsession with both aerial and atmospheric perspectives. He made what we know as his first dated drawing in 1473 when he was twenty-one years old, a picture commonly referred to as the *Arno River Valley*, a view that some art historians once called the first piece of landscape art. He constructed the image from an elevated perspective that makes it feel as if you could step off the ledge at your feet and soar out over the landscape. It's a bird's-eye view initiated from perhaps a specific vantage, but Leonardo has used multiple vanishing points and perspectives almost as if the viewer is in motion above the landscape. It's also useful to note that he has compressed the lateral horizontal plane by moving the hills in more closely and pushing back the horizon so that it provides a very deep theater of action. Its vantage point and pictorial organization underlie what Claude Lorrain would adopt as his compositional strategy, and in turn it thus bears a resemblance to Cole's method.

Leonardo spent time on rooftops and the summit of Tuscany's 1,360-foot Mount Ceceri, Florence visible in the distance, as he contemplated the flights of birds. He wrote that: "The movement of the bird should always be above the clouds, so that the wing does not get wet and to survey more of the landscape . . ." He started drawing flying machines as early as 1481, and worked on their designs until at least 1514. About the view from above he said: "If he seeks valleys, if he wants to disclose great expanses of countryside from the summits of mountains, and if he subsequently wishes to see the horizon of the sea, he is lord of all of them." Leonardo was clearly articulating the idea that to picture a place is to own it, and that seeing it from above is to exercise the ultimate control. This is possibly a fundamental reason why we humans find it useful to continue to exercise our aerial imagination—it's part of the process of converting terrain into territory, a useful geopolitical skill.

In 1502 Cesare Borgia, who was well aware of the power of maps and aerial views, appointed Leonardo as his personal Architect and Chief Engineer, and it was at his request that the artist produced his famous map of the town of Imola, the first mathematically accurate aerial view of a city. Unlike the low oblique view of the Tuscan marshes from 1473, this is a scrupulously accurate dead-straight-down vertical view of the town made by first conducting a geometrical ground survey with compass and measuring tools, and then combining this data set with ground-based observations to construct a rationalized picture. Comparing this view to that of the town of Çatalhöyük isn't to say that the intent of the picture makers was the same, but to acknowledge that some of their cognitive skills were.

During 1503–1504 Leonardo also made a high oblique view of a shallow lake in southern Tuscany, probably in preparation for a strategic draining it of it, which was accomplished only centuries later. Apparently the artist relied on preexisting and somewhat inaccurate maps made by others for much of the typography, but wherever he had himself actually walked, the distances and their scaling from ground truth to map are accurate. And his orography—the depiction of hills and mountains—used shaded relief and specific contours for individual summits, an advance over the stereotypical molehills of the time used to depict terrain.

Carlo Pedretti, perhaps the greatest living da Vinci scholar, has noted that in some of Leonardo's depictions of the horizon he seems to be "observing the world from the airplane," and that in some of the artist's paintings that a bird's-eye view is the organizing principle, whether it was the landscape in the background of the *Mona Lisa*, or the recreations we have of the almost cinematographic lost mural *Battle of Anghiari*. In conversation with Pedretti one evening, he reminded me that Leonardo was an architect, and that my colleague at the Getty Research Institute (and one of his former Ph.D. students) Roberta Panzanelli had written how the artist would "image the ground plan of a painting" before making it. It's as if Leonardo was directing a theater piece and had marked places on the stage for his characters. And that reminded me of how much the view of Çatalhöyük looked like a set design, the map of the town a floor upon which history was conducted, and the volcanoes a scenic backdrop.

Leonardo's aerial views were on the edges of both mapping and painting. His depictions showed the world as it had never been seen before by anyone alive, and they mark the time when cartography would start to drift away from the God's Eye view of Biblical paintings developed in southern Europe to become a secular discipline. By the time Alexander von Humboldt was born in 1769, the two systems of representation, mapping and painting, had separated firmly into horizontally and vertically oriented documents, driven by the need for accurate landscape representations as well as charts.

Alexander von Humboldt, a Prussian aristocrat from Berlin, studied mathematics and geology and took a job as a mine inspector before an inheritance left him independently wealthy, and in 1799 sailed off to South America with his friend the botanical illustrator Aimé Bonpland. Humboldt, who claimed that Leonardo "was the first to start on the road towards the point where all the impressions of our senses converge in the idea of the Unity of Nature," sought to complete that journey. Aaron Sachs has written a splendid book about Humboldt's pervasive influence on the nineteenth century and the invention of environmentalism. He notes that as the scientist's "empirical observations of nature mounted, he grew frustrated with 'mere encyclopedic aggregation.'" Alexander von Humboldt's learned older brother William said his adventuresome sibling was "made to connect ideas." Alexander himself put it this way: "In considering the study of physical phenomena . . . we find its noblest and most important result to be a knowledge of the chain of connection, by which all natural forces are linked together, and made mutually dependent on each other."

And how did Humboldt connect the world? On every continent he traveled—Europe, North and South America, and eventually Asia—he climbed mountains, in particular volcanoes, and carried with him telescopes, sextants, barometers, theodolites, graphometers, dipping compasses, magnetometers, thermometers, electrometers, cyanometers, eudiometers, areometers, and microscopes. As you can see from the frequent occurrence of the word "meter," he was all about quantifying the world, but not just as a matter of listing measurements. By climbing he was able to see enough of the world at one time that he could discern patterns and unifying principles. The most important instance was his discovery, while in South America with Bonpland, that temperatures grew colder as you climbed no matter where you were in the world. He published a map of isotherms showing how the same was true as you gained latitude away from the equator toward the polar regions, and noted that every thousand feet in elevation gained was equal to approximately 150 miles distance from the equator. Furthermore, he then linked ecological communities to his lines of equal temperatures. Go far enough north or south and you would find the same kinds of flora and fauna that you found ascending a mountain in lower latitudes.

Humboldt and Bonpland traveled more than 1,600 miles up the Orinoco and produced the first credible map of the South American interior. They climbed a record nineteen thousand feet up what was then thought to be the tallest mountain in the world, the extinct volcano Chimborazo in Ecuador, and then discovered what we now call the Humboldt Current off the coast of Peru. They continued north and explored Mexico for a year, then sailed to the United States and in 1804 met with Thomas Jefferson, with whom he continued to correspond for years. One result was the founding in 1838 of the U.S. Army Corps of Topographical Engineers, the official exploration unit of the American government. Their goals were exactly those espoused by Humboldt: to go into the field, make precise measurements, record them graphically, and derive rigorously tested unifying theories about what they found. Their most famous member would be John C. Frémont who, as I mentioned while flying over the Great Salt Lake Desert, climbed up to elevated vantage points, and followed exactly those steps to discover the enclosed nature of the Great Basin.

When Humboldt returned to Europe that summer he was the most famous man in the world. Charles Darwin called Humboldt "the greatest scientific traveler who ever lived." Among many other accomplishments, he is credited with initiating the modern field of physical geography, and it never ceases to please me that Denis Cosgrove holds the prestigious Alexander von Humboldt Chair in Geography at UCLA. But Humboldt not only changed the course of science, he also was a profound influence on artists, and in America most particularly on Thomas Cole's promising young student, Frederic Church.

Church was born in Hartford, Connecticut in 1826, and when only eighteen was commended to Cole for study by the older artist's most important patron, Daniel Wadsworth. Cole and Church hiked and sketched together, and one year later the student had his first public showing in New York City. It was 1845, the year the phrase "Manifest Destiny" would first appear in print. Cole died in 1848, and the young Church, who had been hiking through the Berkshires, was ready to assume the role as America's greatest living landscape painter. A key inspiration was publication in English of the book that Humboldt had been writing as the capstone to his career, *Cosmos: A Sketch of a Physical Description of the Universe*. In the second volume of the four-book set, the scientist urged artists to paint nature as if from field sketches, not as the idealized studio landscapes made popular by Claude. The passage reinforced Church's natural inclination, which had already brought him criticism, for painting too realistically—everything from tree bark to rock strata—in his often historically themed canvases. His technique was even declared too photographic, and a bit too topographical for some of the critics.

Church ignored the criticism, and was so enthralled by Humboldt's work that in 1853 he retraced the German's journeys in South America, including ascents up various active volcanoes so he could see their craters and eruptions for himself. In early 1855 he completed the first of his panoramic masterworks, *The Andes of Equador*. He returned to South America in 1857 to climb more peaks, and in 1859 painted the astonishing and encyclopedic *The Heart of the Andes*. Both it and the earlier painting presented the viewer with an elevated position from which to gaze across a landscape that proceeded from tropical jungle upward to snowy peaks. It was as if he had decided to paint the actual progress of the isotherms and their related environments.

Church remained devoted to the heights and extreme places of the world, traveling into Newfoundland and Labrador and even having an

Arctic peak named after him. He had forged a new kind of landscape painting where human history was no longer the governing narrative, but nature itself provided the story. In 1860 he began to purchase what would eventually total 250 acres on a hilltop above the Hudson, and almost directly across the river from Cole's property. He built one of the greatest residences of America, the Orientalist-themed Olana, which he called "the center of the world." Its highest point was an observation tower from which he could look down upon the world and sketch.

What links Leonardo and Humboldt with the Hudson River painters is the idea that only from an elevated vantage point, from the God's Eye perspective, can the world finally be known. You can collect all the measurements you want on the ground below, but in order to understand either human or natural history, you must climb above in order to synthesize the data into a coherent picture. And such knowledge is not just an understanding of how space works, but also time. In order to see change over time you also need to climb above. And that brings me to the contemporary painter Stephen Hannock.

Joe Thompson brings the nose of his red-and-white airplane to the north, swinging us around the Yankee nuclear plant, and we head back toward North Adams. "You have to meet this local guy, the painter Stephen Hannock. I took him for a flight up here and he painted this amazing aerial landscape that's all about the stories on the ground." Sounds good to me, and when we land Joe promises to send me Hannock's e-mail. It always amazes me that you can climb into an airplane, fly over hundreds of square miles in an hour, completely alter your perception of the world— and then drive to the office for a full day's work.

The next day I manage to make an appointment with Hannock to visit him in his North Adams studio, which is located in a former industrial building. Hannock was born in 1951 and grew up near Albany, was a serious hockey jock at Bowdoin College, but loved drawing enough to end up studying with Leonard Baskin at Smith College. He moves with the energy of an ex-athlete, his longish blond-streaked hair never staying on one place about his tanned face. His assistant and colleague David Lachman, an artist who has a studio across the corridor, is fussing with images on a lap-top, and the studio is pleasantly cluttered with a variety of paintings in various states of progression, including large sketches drawn directly on the sheetrock walls.

His most well-known paintings are those based loosely on *The Oxbow* by Thomas Cole, of which there are fourteen versions dating from 1994 through 2001, and shown in places such as the Museum of Fine Art Boston and the Metropolitan Museum of Art in New York, where one hangs next to the original by Cole. Many people, including curators and critics, state that Hannock painted the Oxbow from the same vantage point as Cole, but that's not correct and it unfairly simplifies what Hannock has been doing, as if he were merely copying and updating the scene, which now includes the bridges of I-91 crossing and recrossing the river's northern and southern bends. He is at pains to set the record straight. "I painted that from an imaginary aerial vantage point. There's no foreground—they make me feel claustrophobic. I went to the take-off point for the hang gliders and imagined myself hanging out there a few hundred yards. The omega shape [of the oxbow] is much flatter from the ground. From then on, I've been using this aerial perspective."

Although Hannock uses photographs taken from hang gliders and hel-icopters as informing images, his landscapes are only based loosely upon the actual scene. The mountains in the background of each version of the Oxbow paintings are different, and the degree to which the sun is setting behind the Berkshires varies from painting to painting. As he notes, pho-tographs flatten out everything, whereas he tilts the ground in his land-scapes toward the viewer in order to add depth. The bridges of the interstate look more like aqueducts than in reality, and the edges of the fields are improbably angled, lending energy to the composition. His pic-torial organization has more to do with the traditional bird's-eye perspec-tive handed down from European city-views, than to aerial photographs.

His paintings are usually described as "luminist," their surfaces glow-ing in part because he attacks the various layers with an orbital sander in order to polish them. This three- or four- stage process using increasingly fine grades of wet-dry sandpaper removes the rugosities that would nor-mally bounce light around in haphazard directions, and focuses the reflection directly back at the viewer. He makes yet another connection to the Hudson River School painters by mentioning George Inness, a contemporary of Church, who used a rag to wipe paint from his canvases. Likewise, as Inness matured he moved away from the realism of Cole, Ashley Durand, and Church into a more tonalist and atmospheric style that clearly influences Hannock, and which his sanding promotes.

The subtraction of paint also produces happy accidents. In the first Oxbow version it revealed text that he had inscribed in the underlayers of painting. It was the first time he'd done such a thing, transcribed notes into the work itself, making it a diary of the experiences behind the creation of the image. "In a way it was my answer to Cole, who had inserted himself into his own Oxbow painting." Now it's become a signature of his style. In the Oxbow painting owned by the Boston museum, in which the sun has already set, text furrows the fields and outlines the ridges as if picking up the last of the light. "The vista is an armature for the stories," he says, "a play you can hang light on. Light is what brings you to the stories." In another of his versions of the painting he adopted Cole's inscription of the word *Shaddai*, one of the Hebrew names for God. He placed it on a hillside in the background, and wrote it upside down as if it could be read only from the God's Eye vantage point, as had Cole.

Hanging on one wall of the studio is a study for the painting Hannock did from his flight with Joe. *A Recent History of Art in Western Massachusetts; Flooded River for Lane Faison (Mass MoCA #12)* was painted—one almost wants equally to say written—in 2005. Once again it's a large horizontal landscape painted facing west toward the sun, which balances on the ridgetop behind Williamstown just before setting. A river winds from the center bottom of the painting into the middle distance, and a road spans a bend. Invisible in reproductions of the finished work, but readily apparent in person are the texts, a capsule history of the valley's artists and writers and residents.

"Everything has a Mass MoCA number because we operate in the shadow of it; that's why we moved here, what was happening in the town. And Lane Faison, the art historian at Williams College, because he has been such an important teacher—he taught that all art has a history larger than itself, that it was created during a specific time by someone with a biography. So what are written in the painting are stories about where things are, buzzing a friend's house at the end of the runway, and so on. Then the view is compressed to strengthen composition," which Hannock often does with his aerial panoramas.

The aerial vantage point always leads us to assume that what we're looking at is an accurate transcription of the land, perhaps because it is part of our hard-wired navigational and place-making skills, but Hannock shakes his head at the irony. "The piece I did of the valley here, where Route 2 comes up—it's much wider in reality. I had to narrow it for the composition. Other artists assume, because it's an aerial view, that it's accurate and have copied it." Hannock is continuing to paint his "vistas with text," including one titled *A Recent History of Art in Los Angeles*. His research now includes photos he takes from helicopters and even sketches made from Google Earth.

Finally, I approach the large painting hung on the far wall of the studio. The Oxbow paintings are expansive, dim but warm, and you feel you could just walk into them—in fact, you want to become part of the scene, to be one of the stories. This is an urban scene, however, one set in New York titled *American City with Restored Park (Mass MoCA #10)*. Its palette is more tarnished silver than burnished gold, and you are instinctively not sure you want to intrude. Once again we're suspended over a scene with no foreground, hovering over one end of Madison Square Park and looking across the trees to the Flatiron Building. It's a park that his wife, the late Bridgit Watkins, worked hard to restore before dying of a brain tumor in 2004.

Hannock and his wife were standing at the window of their Greenwich Village apartment on the morning of September 11, 2001 watching the second plane fly into the twin towers of the World Trade Center. Five minutes later the phone rang, a medical lab calling with results of tests run on Bridgit because she had been suffering from double vision during, but then even more worryingly, after her pregnancy with their daughter. The results were not good. That story is in the painting, legible when you get within reading distance. All their stories are there, scratched into the facades of the buildings, a specific period in the lives of two people memorialized from the air, as if the elevated perspective at once made it possible to make them cohere, yet with enough distance for Hannock to handle them. The elegiac palette of the sky, the patterning of the trees in the park, and the language slowly resolving into view as you approach stay indelibly with anyone who's seen the painting. Part of the reason why is that it's an aerial view and tilted toward you, and you can't avoid falling into the picture.

# 9. The 9/11 Trail

*Seen from above, exposition, climax, and denouement all
take place at once. God sees the future as we see the past:
through a trimetragon. In the name of the camera, the film,
and the view itself. Simultaneous eternities are superimposed
to create the illusion of plenitude, but the transposition of
planes is a poor substitute for the transmigration of souls.*
—Ben Lerner

**Joe Thompson banks the red-and-white Tiger to the right, and there**
below me is the ornate bell tower of Frederic Church's Olana, the house
he built above the Hudson. Its arched Moorish windows face the four
winds, making apparent that the real work of the house was to frame the
views around it. The viewshed of the property is considered so historically significant, in fact, that when a cement company proposed in 1999 to
build a large plant nearby, the administrators of the state park hired the
foremost geographical information system software company in the
world, ESRI (Economic and Social Research Institute), to survey it. ESRI
people lofted balloons a hundred feet up from various locations within a
four-mile radius while others attempted to spot them from the house with
both naked eyes and binoculars. The results were plotted with GPS and
used to determine a topographical map of the sightlines. The objective
was to forestall anyone developing structures tall enough to foul the view.
The only substantial impediments I can discover from the plane are the

thousands of trees Church himself planted. But the views south over the lawn and down the Hudson, our flight path this morning, must be magnificent from the tower.

Joe had volunteered to take me on a second aerial tour, this one from North Adams over to Albany, then down over the river to New York City. His father, Jim, is along with us, a retired commercial banker from Norman, Oklahoma who just this last year stopped flying. Joe's wearing a faded strawberry polo shirt, gray cargo pants, sandals and a straw cowboy hat. Jim is crammed into the back seat and wearing shorts and a Mass MoCA hat. Once again we are casual *planeur*, and I run across this attitude time and again with flying families, how they just like to go for a slow cruise around the neighborhood, no particular goal in mind other than to observe the world from above. Jim had owned a single-engine plane of exactly the same model and color scheme as Joe's, and he had first taken his son flying in 1964 when he was six years old. Joe still remembers zooming over the wheatfields and how much he loved it. Joe had offered his dad the controls this morning, but Jim had shaken his head and turned him down. Jim, a tall man like his son, says he was only a recreational flyer. "Never was instrument-rated like Joe is, just a fair-weather flyer." Joe's flown across the country and is much more comfortable with the tight airspace around the city into which we'll be inserting ourselves.

We had pushed the plane out from the hangar a little after 8:00, fueled up, then headed west over the Taconics and to Albany. A cold front is forecast to move down from Canada and be here tomorrow, but as we lifted off there was no more than a slight phototropic haze in the air, the atmosphere here much softer and more golden than the mineral glare of L.A. The luminous light of the Hudson River Valley isn't just an aesthetic construct, but an optical reality.

Joe's flight plan basically followed the flight path of American Airlines Flight 11, the Boeing 767 that was hijacked from Boston's Logan International Airport on the morning of September 11, 2001, and then flown into the North Tower of the World Trade Center in the first of the attacks that day. The hijackers had forced their way into the cockpit of the aircraft, with its eighty-one passengers (including the five terrorists), two pilots, and nine flight attendants, by 8:14 A.M. when they were just east of North

Adams. They had passed south of Mount Greylock at twenty-seven thousand feet at about the same time of morning as we took off. We're not attempting to duplicate their route, but the obvious way to get where we're going follows theirs. I had been unsure how I could write about flying in this part of the country, especially when dealing with aerial views of New York City, without confronting the events of 9/11. Joe's offer made them unavoidable.

As we had crossed the Taconic Range, the skyline of Albany had immediately come into view with its distinctive and singular downtown office tower, a landmark for the hijackers as well as for us. Albany itself is a geographical and historical marker, not just as the state capitol, but also where the waters of the Hudson change course. Just north of the city is the Federal Dam, the first impediment to navigation on the river 153 miles or so up from Battery Park on the lower end of Manhattan. Boats use locks on the east end of the dam to enter the Mohawk River and the Erie Canal. And from Albany heading north it's less than 180 miles to Canada, not just a different countryside but a different country.

We crossed reservoirs, ponds, creeks, every kind of inland water body known, then upon reaching the river turned left and southward. Flight 11 had kept going straight for about twenty miles before the hijacker could redirect the momentum of the fast-moving jetliner. The valley below the state capitol is broad and well settled, and holds a major commercial thoroughfare, the passenger trains taking the east bank to our left, freight trains on the west bank to our right. In fact, the shoreline of the Hudson from here to Manhattan Island is more an engineered roadbed for the railways than a natural feature. The reasons why become obvious from the air. The sublime scenery of the Catskills is more than compensated for by the almost continuous aggregate of generator stations, cement plants, quarries, old brickyards, and manufacturing sites for chemicals, plastics, and pharmaceuticals.

It looked like the tide was out. A high of tide 5.16 feet was due at 3:51 this morning, the low tide down more than six feet at 10:42 A.M., one of the two daily pulses that brings saltwater nearly halfway up to Albany from the Atlantic Ocean. From there it was seventy miles south to Olana. Out the window to the right were the buildings of the Watervliet Arsenal. The first buildings on the 150-acre site were built in 1813 as a defense against the British coming south from Canada during the War of 1812, which makes it the oldest continuously operated weapons factory site in the country. It remains the only arsenal making large-bore cannon, which they advertise as the finest in the world. Laid out side-by-side on the front lawn are enormous tapered cylinders that are battleship gun barrels. The M1 Abrams tank barrels are also made here, but are small enough to be kept inside and out of sight as we pass.

The power plants, limestone quarries, and cement plants alternate as we go downriver, and if it weren't for the fact that the Center for Land Use Interpretation had recently bought a building just upriver in Troy as an East Coast facility, and was thus compiling a database on the industrial and military sites in the area, I'd never be able to keep track of them. I'd say it's coincidental that CLUI and I end up working in the same places, except you can't talk about the nature of space and place in the contemporary world without continually engaging the military-industrial complex, which the CLUI mission statement obliges them to document. The power industry, construction materials, and military manufacturing are here because of the river as a route for transporting heavy materials, be it cement, coal, fuel, or steel. But also because it was for decades a convenient place in which to dump the toxic byproducts.

The last major site on the river that I'd made a note of before reaching Olana was not far upstream on the opposite bank, a mile-long covered conveyor belt leading from a cement plant to the west and ending at the river's edge where it brought product to barges. Other lines ran even farther west to the limestone quarries. The plant is owned by LaFarge, the largest construction material company in the world, and since 2001 the largest cement producer. The modern French company started out in 1883 with the purchase of a local lime producer that had been around since 1749. By 1864 the company was supplying lime by the hundreds of thousands of tons for the construction of the Suez Canal. From 1991 to 2001 the demand for cement in the United States increased 50 percent, and it continues to grow, only now in competition with the Chinese construction industry. And yes, LaFarge is in that country, too.

We were almost thirty-five miles south of Albany when Joe spotted the top of Olana sticking up above the trees, positioned on a high hill with what would once have been a 360° view, not least of which was across the river to the Catskills and Thomas Cole's Cedar Grove property. It gave Frederic Church great satisfaction to be able to see the literal space in which his personal history as a painter developed, as well as that of the entire Hudson River School. It would have been a shame had St. Lawrence Cement been allowed to build its plant in the nearby town of Hudson. It was proposed as a coal-fired industrial complex with a 1,200-acre open-pit mine, forty acres of buildings, and a four-hundred-foot smokestack that would have produced a plume of smoke as long as six miles. A two-mile-long conveyor belt would have connected the plant to the waterfront where barges and trucks would have been swapping fly ash and fuel with cement and heavy-metal waste. The Olana people with their aerial survey were, however, just part of a larger environmental coalition that over the years has begun to clean up the river as many heavy industries phase out and move elsewhere.

We leave Olana, now at about the spot where Flight 11 rejoined the river, and although the jetliner had been descending here, it was still flying at an altitude of about fifteen thousand feet, far above our position. A few minutes later and on the east bank, about seventy miles south of Albany and ninety miles north of the city, a mansion looks out over the water. It's the Vanderbilt place, one of the huge estates built on that side of the river at the turn of the twentieth century in order to enjoy some of the scenery that the painters had been selling from the New York galleries to earlier generations of the family. The fifty-room house faced with limestone in 1898 had a year-round staff of sixty, although the family used it for only a few weeks in spring and fall. After all, the grandchildren of the railroad tycoon had other homes to enjoy, including the larger property at Newport and the mansion on Fifth Avenue.

We pass by Poughkeepsie and another huge quarry, and downstream I recognize the distinctive rooflines and skylights of the Dia Art Foundation's Beacon facility, another factory that's been turned into a contemporary exhibition space as tourism fills in the structures abandoned by industry. The river bends around the 1,340-foot Storm King Mountain and there's West Point, which has been occupied since 1778 by the military. Even from the air its gray neo-gothic buildings are stern and foreboding.

The river is at its narrowest and deepest here, 202 feet on average, depending on the tides. We're flying below the bluff tops now and for the first time the city has come in sight, downtown Manhattan just a ridge of buildings in the distance.

Joe pushes forward on the controls and the red-and-white Grumman Tiger begins to lose altitude. We're about forty miles out of New York City, flying south along the Hudson River past West Point, and coming up on the Indian Point nuclear power plant, which is very much still in operation. Joe is careful to stay across the river from it, given the post-9/11 flight rules. The hijackers of the airplanes had apparently considered trying to take out the containment dome and reactor, and for some reason decided instead to aim at the World Trade Center. High clouds are drifting in from the west and there's a light scrim of haze, but the outline of mid-Manhattan is already breaking into a fractal profile of individual buildings. Indian Point generates two thousand megawatts and is the largest single-point source of electricity for the city, about 25 percent. If they had successfully hit the reactor and broken it open, the resulting loss of power and airborne radioactive contamination could have affected up to 7 percent of the entire U.S. population. It's now one of the most heavily protected power plants in the world.

Joe glances down at the chart book in his lap. "We're flying into a tightly layered airspace. When we cross the Tappan Zee Bridge, the beginning of the controlled airspace, we have to stay under three thousand feet. Over the Washington Bridge we have to drop to under eleven hundred feet. Around the Statue of Liberty we'll be at about eight hundred feet or so, then when we turn left to go up along Long Island, we'll have to be at five hundred feet or less to get underneath the traffic pattern for JFK. It would be kinda embarrassing to be forced down by F-16s."

Earlier in the year with Michael Light over the great arid spaces of the West, I'd been writing about the military presence that is mostly invisible from the ground, yet all too apparent from the air. Mike and I were cruising in between restricted airspaces to peer at sites such as the Dugway Proving Ground, where we developed napalm and techniques for firestorming cities in World War II. Just as with Joe today, Mike flew with a chart book handy and kept very clear in his head where you stroll and don't.

General aviation has always been and remains a libertarian pastime in that it's mostly unregulated if you stay out of the commercial air traffic corridors and military airspace. Airport security screening for private pilots and their passengers is almost nonexistent, and much to my astonishment we remain free to fly over major ports, oil refineries, and bridges. And, in fact, as we approach the George Washington Bridge, we sink below a thousand feet. We're lower than the top of the Empire State Building; the Manhattan skyline is no longer a horizon line, but a wall of architecture in front of us. On an earlier flight, Joe had quoted the artist-cum-pilot James Turrell about what it's like to fly just below one thousand feet, where the horizon is around the halfway point on the windscreen, and you're "in the nap of the land." He should know.

Out West, Mike and I were flying over military history, but everything was abstracted, old target patterns plowed into the desert floor like enormous bull's-eyes. They could be mistaken for art. But here, coming down to the level of the river bluffs and just a couple hundred feet above the towers of the bridge, the faces of drivers visible in the cars crossing its 3,500-foot suspension span—here we're within the traffic pattern. And in the land. And in history. Our flight vector has now intercepted that of the hijacked plane and we're at the altitude at which Flight 11 flew into the windows of the North Tower. Flight attendant Amy Sweeney made the last transmissions from the plane, speaking on her cell phone from this same airspace. Asked by flight controller Michael Woodward to describe what she saw, she replied "buildings," and then "water." Then, as they approached the bridge: "We are flying low. We are flying very, very low."

It's here, even before we fly by Ground Zero, that 9/11 stops being a date for me and is suddenly part of my own life, my own trajectory. There's nothing to see—no jetliner, no towers—and nothing to hear save the drone of the 180-horsepower engine and wind. And yet here is this other person looking out the window, watching the bridge go by, and her last words: "Oh my God, we are way too low." She knew what the implications were. The jetliner was flying at 470 mph, more than four times our speed, and within two minutes had plowed into its target.

The aerial has always been a viewpoint for me, a vantage point through which I moved when my attention lingered long enough on the ground below to establish a relationship between the ground and my brain. What I would call a place has usually been the space *toward* which I have gazed, my vision colonizing it to my purpose. But now the point *from* which I am viewing an event becomes a place. Airspace becomes airplace.

Flying down the west shore of Manhattan Island, all I can do is stare out the windscreen. It's impossible for me to take notes. There's the density of the city, but all I can think about is the hole at the lower end of the island, where the fractal massing of the office buildings builds up to a point, a summit, that's no longer there.

The top of the Empire State Building shines in the sun. A military helicopter zooms past on our left side headed upriver, flying in between us and the skyscrapers. The gray and aging *USS Intrepid* aircraft carrier and *Growler* submarine are docked below, relics of World War II and the Cold War. As will happen when viewing the earth from above, a mental circuit trips and I connect those military sites in the West to the attack on what was considered our tallest mosque, our supreme temple of commerce. We are always preparing for war, which is an invitation for it to come visiting.

# 10. The Vertical City

*She has taped an aerial photograph of our neighborhood to
the ceiling. She looks up to see our house from above while
we're in bed. This is but one example of her uncontrollable
desire to look down on the structures that she's in.*

*Photographed from above, the shadows of the soldiers seem
to stand upright, casting bodies. Birds are rarely depicted
from a bird's-eye view. From this angle, she doesn't love me.*
—Ben Lerner

**Michael Richards was at his studio on the ninety-second floor of the**
North Tower on the morning of September 11th, a low-ceilinged room
with narrow windows facing the Statue of Liberty. He was one of the
twenty-five artists working in the Lower Manhattan Cultural Council's
"Studio in the Sky" program, which since 1997 had been hosting artists in
otherwise unoccupied office space in the World Trade Center. Richards
had called his girlfriend around midnight to say that he was going to
work a bit, then spend the night in the studio before heading over to the
Bronx Museum at ten, where he was employed as an art handler. At 8:46
the hijacked American Airlines Flight 11 slammed into floors 93 through
99. He would have just been getting ready to leave. Michael Richard's
"remains" were identified six days later.

Richards, a thirty-eight-year-old Jamaica-born artist who had earned his master's at New York University, had been working on the figure of a man falling to earth in a storm of flaming debris, one of a series about the "Tuskegee Airmen" of World War II. The African-American 332nd Fighter Group had been formed in 1941 as part of the U.S. Army Air Corps, and more than a thousand personnel trained during the war at their base in Alabama. More than four hundred of them had been deployed to North Africa and Europe, where they flew fifteen thousand sorties and earned 150 Distinguished Flying Crosses. Like many minorities who have fought for our country during various wars, they didn't receive much in the way of acclaim for their sacrifice at the time. After 1945 the group continued to fight discrimination in the military, eventually helping the U.S. Air Force to integrate its forces. In 2007 approximately 350 of the Tuskegee Airmen were finally and collectively awarded the Congressional Gold Medal in recognition of their service.

Richards had been creating sculptures of and related to the airmen for eight years, using his own body for the mold and casting the figures in bronze-plated resin. In an earlier exhibition he had placed the upper torsos of three airmen in a plaza as if they had fallen from the sky and plunged into the ground feet first. Another piece in the series was titled *Tar Baby vs. St. Sebastian*, an aviator in World War II flying garb pierced from the waist upwards by small airplanes, an allusion to the arrows impaling the third-century martyr who was killed for refusing to denounce his faith. Richards said of his series that he was honoring men who flew heroically even as they were dragged back to earth by prejudice. To say that his work was ironically and horribly prophetic barely begins to address the fact that nearly 250 people jumped or fell from the burning buildings of the World Trade Center.

Vanessa Lawrence was in the building that morning, too, getting there early to catch first morning light for her panoramic painting of the city to the north. She had gone downstairs to make a phone call, and had taken the elevator back to her studio space on the 91st floor when, just as she stepped out, the building was suddenly shaken with a tremendous force. She headed for the emergency stairs and during the next hour made her way down to street level. She ran across the plaza where a firefighter covered her with his jacket as the South Tower came down. A half hour later the North Tower followed.

When I wrote in the last chapter that the World Trade Center (WTC) was our tallest mosque to commerce, I was alluding to its metaphorical attractiveness as a target, but also its literal architecture and aesthetic roots. When site demolition and groundbreaking began on the sixteen-acre plot in 1966, the 1,360-foot-plus towers were meant to be the tallest buildings in the world. Tenants began moving into Tower One (the North Tower) in late 1970, and two years later into Tower Two. The buildings held almost a square acre of rentable space per floor built around the load-bearing central core, and although it took until the 1990s for the buildings to begin to fill, eventually they accommodated up to fifty thousand people, which the artist Olu Oguibe points out was the population of a decent-sized Medieval city. His observation is relevant to the design of the WTC, as it was during the Middle Ages that Europe began to import pointed arches into its cathedrals from Moorish designs. This melding of holy architectures resonates with the contemporary use of Islamic motifs in what was to be a temple of commerce.

The architect for the WTC was Minoru Yamasaki, a Japanese-American who in the late 1950s was commissioned by the Saudis to design the airport terminal for Dhahran. He gave the facility a long facade of pointed arches, a minaret for an air control tower, and prefabricated concrete forms that resembled the traditional tracery of Islamic art and architecture. The Saudis liked the building enough to reproduce an image of it on a banknote, and Yamasaki was so taken with his Islamic modernism that he used it in numerous other projects—including the World Trade Center, which he was commissioned to design the year after the Dhahran Airport was finished. Laurie Kerr, writing in the online magazine *Slate* in late December 2001, reported on how the architect deliberately echoed the plan and features of Mecca's courtyard with its two minarets on the New York site. The pointed arches at the base of the towers, and the filigree of the exterior truss of the buildings, were overt references to traditional Islamic architecture. Bin Laden and his fellow terrorists may have had other reasons to attack the WTC, among them its population density and even the fact that the corrupt Saudi royal family was a patron of Yamasaki. But if nothing else, Bin Laden was aiming to take down the tallest minarets of his enemy.

Strip away the Islamic references, however, and you were left with the ultimate expression of the architectural urban modernity and its attendant hubris. The tall minimalist structures were in theory cost-effective, but they completely ignored and dominated the scale of the surrounding built environment of Lower Manhattan. Whatever cultural allusions Yamasaki deployed, the size of the towers overwhelmed them and the banal buildings stood mostly empty for ten years. It took until the 1980s and the construction of several adjacent buildings for the neighborhood to begin to grow up to the scale of the towers, to create a pyramidal mass in which the towers could make sense as a monument. If at first the WTC stuck out like a sore thumb from the much lower built horizon of Lower Manhattan, as the small buildings stepped up to them, they created a physical investment in and corollary to the growing emotional attachment of locals. When the towers collapsed that pattern was destroyed, creating a profound rupture in the urban texture that is never more apparent than when seen from the air.

If the towers at first posed a problem of visual scale for local residents, they also presented potential office workers with another problem related to their height, which was vertigo, a condition that almost comically afflicted Yamasaki. He designed the majority of the 43,600 windows to be only eighteen inches across, about the width of an average person, to provide a sense of safety. As the architectural historian Anthony Vidler points out, other measures included deep and intimate office spaces so that people could retreat from the dizzying views.

From the observation decks near the tops of the South Tower you could see forty-five miles in every direction, and that proved an irresistible lure for artists from around the world. Yvonne Jacquette, a painter noted for making aerial views from airplanes, started working from the observation deck almost as soon as it opened, and in 1996 the Manhattan-born painter Carl Scorza was looking for a window in the towers from which to paint the city. The result was the World Views residency program, which in 1997 began offering empty office spaces to artists as studios. The initial group of eighteen artists, Scorza among them, occupied unimproved spaces on several floors from the 24th up to the 91st, but in 1998 the program was re-conceived as an artist's colony for the 91st floor of the North

Tower. By 9/11 more than 130 artists had participated in the program. The fact that Scorza went on to spend two months in 2000 in residence at Olana embodies the link between the Hudson River artists of the nineteenth century painting their elevated views of a sublime landscape, thus representing the start of Manifest Destiny, and the WTC artists painting the sublime canyons of New York City as the result of that capitalist expansionism.

Urban fabric is created physically, at first by the presence of buildings, but also mentally by the visual representations of them that come to exist over time in maps and photographs, in films and paintings. Tall buildings always attract image makers, and once the Studio in the Sky program was announced, there was no lack of applicants who wanted access to the views, who wanted to weave part of that fabric. Out of the eighteen participants in "Pilot Project: The Perceptual Painters" held in 1997–1998, all of them save Rackstraw Downes (who painted wonderfully moody interior panoramas of unrented interior spaces) made views out the windows. The artist Susanna Heller said she felt as if she was flying through space when in her studio, and Elisa Jensen noted she was painting from a vantage point that had never existed before.

While Jacquette made nighttime views that were as close to being aimed vertically down as she could get, a corollary to her views painted as if from airplane windows, Ron Milewic sent his gaze outward over that forty-five miles and painted mostly skies unscored by the Manhattan skyline. As time went on, fewer artists looked out the windows—who wanted to repeat what the others had done?—and worked more with interior installation. But at least 26 of the more than 130 artists who would inhabit the buildings continued the aerial tradition. What is remarkable is that so many of their views are the same and yet not. If they painted the view to the left, it was the Hudson River and New Jersey; if to the center, as with Vanessa Lawrence, the Empire State Building was the focal point; if to the right, it was the Brooklyn Bridge and the narrow corridor of the East River.

Daniel Kohn and Sonya Sklaroff used the skinny vertical format of the windows to frame views, and Mary Jane Dean stepped back from the sills to photograph the windows at night with a pinhole camera. The mullions are almost entirely black, while through seven slits the city outside is washed in light. The photos look as if they were printed in reverse and remind me of David Maisel's *Oblivion* pictures. Joellyn Duesberry, who felt as if she had spent months "up in the air," was fascinated by the cubical geometry of the city. In a wry acknowledgment of how the artists were reinforcing the iconic height of the towers, Robert Selwyn photographed paintings of aerial views of the WTC. The vantage points from the tower were limited in number and severely constrained, yet the views, the pictures, were never the same.

If in Los Angeles artists have emphasized the horizontal grid of the city, in New York the typical view is looking down on it from within it, from one of its own buildings. This particular framing of an urban view has its roots in both the history of photography and the European cities where the technology was first used. Niépce was photographing from his third-story window in 1826 as he made the first photograph, in part out of necessity: it was the only place in the house where he could get enough light over the course of hours necessary to form an image. When Daguerre announced his own version of the process in 1839, he had already been making ten-minute-long views from a window overlooking the Boulevard du Temple. His first outdoor use of his apparatus in 1841 was a view of the Pont-Neuf from an elevated vantage point above the Seine. In 1842 Antoine F. J. Claudet took a 360° panoramic series of photos of London from the top of the Duke of York Column in Pall Mall. These were all images of cities that were predominantly horizontal. The Eiffel Tower was the harbinger of profound change in the profiles of cities, and consequently their images.

From the moment construction on the Eiffel Tower began in 1887 photographers were clambering up its steel beams; when it opened in 1889 it was the tallest structure in the world and would remain so until the Chrysler Building was built in 1930. Alphonse Liébert, who had been working from balloons above Paris for more than a decade, took a photograph of it from the air as a completed structure during 1889 Paris Exposition. The links between the Eiffel Tower, New York, elevated city views, and aereality are so numerous and interesting that they deserve an investigation all of their own, but we can tease out just a few threads here to see how important they are to the history of aerial imagery.

Prior to the late nineteenth century, New York had, like Paris, been a horizontal city of buildings that for the most part reached no higher than four or five stories, about the limit for residencies and offices served by stairs. You could load masonry with vertical weight up to almost two hundred feet, but anything over eight stories and the bottom walls had to be six feet thick to support it. The invention of the skyscraper, which used steel and then later reinforced concrete as a load-bearing skeleton, allowed the profile of cities such as Chicago and New York to move aggressively upward. And the artists followed. The first steel building in New York was the since-demolished Tower Building on Broadway built in 1888–1889, at eleven stories and 130 feet, a relatively modest project. The Park Row (1896–1899) went up 30 stories, and photographers wasted no time in ascending the city's tallest building.

The Flatiron Building, finished in 1902, was shorter than Park Row, at only 22 stories and 311 feet; but, unlike the earlier building, which received almost unanimous scorn for its stolid appearance, the dramatic prow of the Flatiron was a showstopper. Built at the triangular intersection of 23rd Street, Fifth Avenue, and Broadway (and overlooking Madison Square Park, per the painting by Hannock), it was instantly a popular place from which to make elevated views of the city. The unusual openness of the intersection provided ample light for photographers, and created diverging points of perspective that reinforced the dynamic nature of the city. The Singer Tower went up six years later at almost double that height, however, and the race for the tallest building in the world got up a serious head of steam.

Most of the images taken from those heights were journalistic by nature, more interested in promoting the vigor of a pre-crash economy than anything else, but Georgia O'Keeffe and her husband Alfred Stieglitz would soon participate in the formation of an aerial aesthetic for New York. Stieglitz, born in New York in 1864, made one of the most famous photographic images of the early twentieth century in 1903, that of a steam engine in winter chuffing through the New York Central Yards. The picture was made from high in a building. Stieglitz photographed out of hotel and apartment windows in both Berlin and New York in 1907, and looked up to catch a portrait of a biplane in 1910. He first saw O'Keeffe's drawings in January of 1916, gave her a show that spring, and in 1918 offered to support her in New York while she worked. The rest is art history.

The two of them started spending each summer and fall at the Stieglitz retreat on Lake George. Aerial views of the resort areas had become popular after World War I—another result of the aviators returning home— and such images may have helped inspire O'Keeffe when in 1921 she painted *Lake George with Crows*, her first use of a genuinely aerial perspective. That same autumn Stieglitz started to photograph clouds with his Graflex 4x5. The couple was friends with every notable artist in New York, including the modernist painter who would influence Margaret Bourke-White, Charles Sheeler. Sheeler was photographing the city from rooftops in 1920 and O'Keeffe sketching the same views the next year. She and Stieglitz married in 1924 and moved into rooms at the Shelton Hotel on Lexington between 48th and 49th, in November of 1925. For twelve years they lived there on the 30th floor where they were high enough to get good north light for painting, but also great views to the east. From 1926 to 1929 O'Keeffe worked on a series of paintings from the windows of their apartment and those of neighbors, many of them over the East River and framed almost identically to the photos taken by Stieglitz from the same vantage point with a telephoto lens. "Yes," she said, "I realize it's unusual for an artist to want to work way up near the roof of a big hotel . . . but . . . he has to have a place where he can behold the city as a unit before his eyes but at the same time have enough space left to work."

Then there was André Kertész, the Hungarian photographer born in 1894 who had a habit of climbing church spires and mountains. As early as 1919 he was leaning out of apartment windows in Budapest to take photographs of the sidewalk life below him, and the first morning he woke up in Paris upon emigrating there in 1925 he did the same thing, sticking a camera out of his hotel window. Kertész hung out with the dadists, as well as Chagall and Calder, the writer Colette and the painter Piet Mondrian, who was his senior by more than two decades. Influenced by the abstract Constructivist paintings and the modernist work of the

Bauhaus, in particular that of its leader and his fellow countryman Moholy-Nagy, he became an important magazine photographer in Paris in the 1920s and '30s. His most important project from the period was his second book, *Paris Vu par André Kertész* (1934). Its title page image is a photo of thirteen pedestrians crossing an intersection below in diagonals, from a curb at the lower right headed toward the upper left, a line continued across street by the walkers. This is not a photograph about the people—they are completely anonymous—but an important formalist scheme he would use throughout his life. It echoes one he did on assignment in 1929 for the fortieth anniversary of the most famous monument in Paris. *Shadows of the Eiffel Tower* was taken looking down from the observation deck on the first level almost 190 feet up, and onto pedestrians crossing the plaza beneath its legs.

The Keystone Photo Service invited Kertész to New York City on a one-year contract in 1936. It was a poor fit, but there he was that year making a self-portrait as he leaned out a window of the Hotel Beaux-Arts on 44th Street. His photographs of the city look up from windows as well as down, and remind us that O'Keeffe had painted a stunning view of her residence, *The Shelton with Sunspots*, while looking upward from street level in 1926. Living in and amongst tall buildings allowed, or perhaps forced, residents of cities to mentally manipulate space in new ways, and the artworks manifested the change.

Kertész was hired by *Life* in the late 1930s to photograph the waterfronts of New York and New Jersey. Seeking a way to make the familiar strange, he hired a dirigible to take him around the Empire State Building, which he shot as a high oblique, and then over to the Hudson River, where he captured the rail yards at a more moderate angle. He also photographed the assignment on the ground and in tugboats. Weston Naef calls it "one of the most accomplished essays in the early history of American photojournalism," but the magazine, sadly, never ran it.

In 1952 Kertész moved to a terrace apartment on Fifth Avenue, a place on the twelfth floor with a balcony overlooking Washington Square Park in Greenwich Village. He prowled the floors while the building was under construction, looking for exactly the right vantage point. It was one of the few relatively tall buildings at the time being built in Lower Manhattan, an area that reminded him of Paris with its mix of residential and commercial buildings. Kertész worked from that balcony until his death in 1985. He used several lenses of different focal lengths, allowing him to frame and compose selectively from his perch, and he made hundreds of pictures of the square below. His image of a solitary walker in snow among bare trees is an iconic view of the city, elegant and carefully constructed, a precise echo of a 1935 picture he made in Paris, *The Vert-Galant under the Snow*. In all pictures from above the people remain anonymous, just as they were under the Eiffel Tower, and yet the photographs have an intimacy akin to that of those taken by Terry Evans over Chicago. His altitude, like hers, was in that zone where, as he put it, everything was both high and low

In 1972 Kertész took a photograph of the newest skyscrapers in his part of the city, the World Trade Center. It's a foggy day, the upper reaches of the Twin Towers lost in cloud. The bell tower of a church is in the foreground, a dark cross standing just to the left of the silvery gap in between the two towers. Below the fog level and just to the outer edge of the right-hand tower an airplane is about to circle in front of the buildings. It was not a well-known image by the artist, but it was resuscitated after 9/11.

Berenice Abbott was younger than these artists, born in 1898. She moved to New York City in 1918 to be a journalist, fell under the sway of Man Ray and other cultural figures, and ended up in Paris studying sculpture. In 1923 Man Ray asked her if she would assist him in his Montparnesse studio, and once she experienced how images floated up out of the developing solution, she never looked back. She had her first solo show in 1926, exhibited with Kertész in 1928, and then the next year sailed for New York to look for a publisher interested in the photographs by her recently deceased friend Eugene Atget. She was stunned at the vertical transformation of the city. The Chrysler and Empire State buildings were vying for first place as the tallest buildings in the world, and literally a thousand tall buildings crowded the sky in the financial district. She closed the Paris studio, got herself a glass-plate 8 x 10 large format camera, and started leaning out of windows.

Her photo *Wall Street District*, with the street far below leading from lower left to upper right, that traditional aerial diagonal, is photographed with a wide lens from the rooftop of the fifty-five-story One Wall Street tower. All the buildings seem to lean outward from her in a composition that is at once dizzying and exhilarating and dismaying, a well-known early twentieth-century trope for life on the street at the feet of the sky-scrapers. Her photograph of a temporary parking space far below for the Rockefeller Center—unlike the Los Angeles ones made for Ruscha—is filled with cars. Once again she tilts the camera to achieve that same diag-onal line, a way to keep the elevated view from become static, and a tech-nique familiar to Mike Light while working over L.A.

Abbott didn't particularly like heights, unlike Bourke-White who was working in the city at the same time (although Abbott admired her work). Most of the views she made of New York were taken from the ground, including all but 20 of the 308 in her project *Changing New York*, which was sponsored by the Federal Art Project of the Works Progress Administra-tion and published in 1939. The twenty elevated views are mostly hand-some portraits of skyscrapers in the Financial District and the avenues in between them. The other 288 have their vantage point anchored res-olutely to the ground, their often steep upward tilt capturing the tops of the buildings, making them lean inwards as if the city is about to fall on us. Berenice Abbott may have loved New York, but she knew the effect the canyons created by the tall building could have on your psyche. It's a bit ironic, therefore, although perhaps not totally surprising, that her most famous view of the city was a reversal of this visual field.

One night in 1932 she leaned out over the edge of the Empire State Building, when such a thing was still possible, and made *New York at Night, Empire State Building, 350 Fifth Avenue, West Side, 34th and 33rd Streets (General View North), Manhattan.* The title is more a map than art, and it's often shortened to *Night View, New York* or some variation thereof. The picture is a stunning black-and-white mass of office windows lit far below in the darkness, the light so dense that the night seems all but ban-ished. It seems impossible to have been a view made from a building, you are so high in the air. And it is this photograph that Ed Ruscha bought for his personal collection, the long exposure of a fabled city lit by electricity, a vertical corollary to his horizontal night views of Los Angeles.

People will climb up to the highest vantage points they can find in order to get a grasp of where they live. Children pull themselves up into trees and

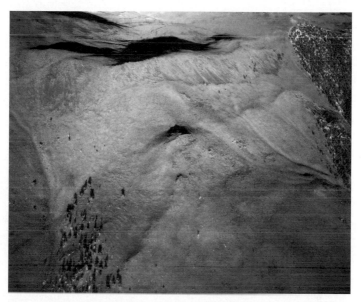

Plate 1: Michael Light, *Bristlecone Pines at 1000' 1830 hours, White Mountains, CA,* 2001. 40 x 50 inches, monochrome photograph.

Plate 2: Michael Light, *Tailings of Barney's Canyon Gold Mine, Near Bingham Canyon, Looking North, UT,* 2006. 40 x 50 inches, monochrome photograph.

Plate 3: Michael Light, *1215-Foot-Tall Garfield Copper Smelting Stack, Great Salt Lake Beyond, UT*, 2006. 40 x 50 inches, monochrome photograph.

Plate 4: Michael Light, *1215-Foot-Tall Garfield Copper Smelting Stack, Great Salt Lake Beyond, UT,* 2006. 40 x 50 inches, monochrome photograph.

Plate 5: David Maisel, *Oblivion 15n*, 2004. 40 x 40 inches, black-and-white photograph.

Plate 6: Terry Evans, *Ammunition Storage Bunkers and Hay Bales,* 1995. Color photograph.

Plate 7: Eugene von Guerard, *North-east view from the northern top of Mount Kosciusko* [sic], 1863. 36 1/5 x 54 1/3 inches, oil on canvas.

Plate 8: Mandy Martin, *Wanderers in the Desert of the Real, the Tailings Dam*, 2008. 70 3/4 x 106 1/3 inches, oil, ochres, and pigment on linen.

Plate 9: Fred Williams, *Winjana Gorge, Kimberley, I*, 1981. 59 7/8 x 72 inches, oil on canvas.

Plate 10: Michael Light, *Untitled/Sunset Boulevard Los Angeles*, 2007. 40 x 50 inches, monochrome photograph.

onto the roofs of their homes, apartment buildings, water towers on ranches—and anything else that's within reach. The views made by artists from apartment windows and balconies are part and parcel of the endeavor to humanize the urban environment by reconceiving it into patterns we can understand. The Eiffel Tower was lauded as offering for the first time an opportunity for Parisians to encompass enough of their city at one time so that they could understand it as a unified whole, much the same objective as that held by the makers of bird's-eye views of Venice centuries earlier. The same was said of the Empire State Building when its observation deck opened in 1931. As historian David Nye described in "The Sublime and the Skyline," his well-known essay on New York's skyscrapers and the creation of the geometrical sublime: "The new vantage point seemed to empower a visitor, inverting the sense of insignificance that skyscrapers could induce when seen from the ground. The observation platform offered a reconception of urban space, miniaturizing the city into a pattern."

"Miniaturizing the city into a pattern" is, of course, exactly what children are doing when driving toy cars around streets pictured in an aerial photomap. Looking down on a city from on high is a way of manipulating the space we live into the pattern of a place, thus giving us a measure of control over at least our response to it. Roland Barthes, as quoted in that same essay by Nye, proposed that the Eiffel Tower overlooked "not nature, but the city; and yet, by its very position as a visited overlook, the Tower makes the city into a kind of nature; it constitutes the swarming of men into a landscape." And that process of looking over brings the city into a realm where our neurophysiology can process it and create the kind of paradoxes about power that attract cultural critics. But just as people scale the highest summit the city has to offer, and artists both use it to picture the city and subvert our notions about it, the airplane has offered a way to put even that exalted vantage point into perspective. Margaret Bourke-White made her photograph from an airplane looking down on an airplane above the Empire State Building. And it wasn't just photographers working from the skies, but also painters.

The British artist C. R. W. Nevinson started out as a postimpressionist painter, but in 1912 came under the influence of the Italian millionaire poet and founder of the Futurist movement, Filippo Tommaso Marinetti.

The movement glorified speed, the internal combustion engine, and destruction, and airplanes were an important icon in the vocabulary of the poets and artists associated with it. Marinetti had taken Italian artists to Paris to expose them to cubism. The fact that Picasso painted the world's greatest anti-fascism paintings, *Guernica*, which was based on the aerial bombing of the Basque town, but that the Futurists would be inspired by cubism to promote bombing as a supremely creative act, is one of the sadder ironies in art history.

Nevinson coauthored a Futurist Manifesto for English Art with Marinetti in 1914, and adopted the fractured geometry of cubism; but when he rejected Marinetti's claims that war was "the world's only hygiene," the two parted ways. In 1915 Nevinson made an early attempt to picture aerial combat with searchlights at night carving up the sky during a raid. It wasn't until 1917 that he actually got into the skies. He had first asked to be assigned to the camouflage section where he could put his painting skills to good use, but ended up as an official war artist flying over German lines as an observer first in a balloon, then later in a reconnaissance plane. He was chased by fighter planes, and at one point had to parachute out to safety. His views of aerial combat, such as *Spiral Descent* (1916) with a biplane creating a vortex in air displayed both a realistic grasp of how aerodynamics altered human perception, as well as how he bent the futurist influence to his own ends. He painted views of both Britain and France from the air, the geometry of the fields an appealing subject even as war raged in the skies above them.

After the war Nevinson wanted to try his hand in the art market of New York. He didn't make much of a splash and shortly returned to London, but subsequently created a series of paintings and prints of the city in 1919 and 1920, many of them views from windows on high, titling one view of skyscrapers from his apartment *The Temples of New York*. His graphic use of a cubist-inflected aerial perspective influenced many artists in the city, among them the young Georgia O'Keeffe. The apotheosis of the aerial an increasing abstraction would of course be Piet Mondrian's city grids.

It is within the European tradition of photographing from windows that Yvonne Jacquette began painting aerial views of cities, although she was using the windows of an airplane to do so, making bird's-eye views that were based on a real aerial experience versus an imaginary one. Yet, her paintings arc only topographical in the broadest sense. Jacquette was born in 1934, attending the Rhode Island School of Design from 1952 to 1955, and took a job after school drafting illustrations for helicopter manuals. She was a fine realist painter, but always maintained an interest in abstraction as a way of forming a more generalized view of the world. In 1967 she was painting a picture of the interior of a barn, then three years later making an oil-on-canvas of the clouds out one of its windows. Like O'Keeffe, she would find herself painting views of New York City from the street looking up at night, indications to me that she was intrigued by the ability to manipulate space mentally, to rotate it into eccentric yet revealing perspectives.

In 1971 she took a plane ride to San Diego and started to make water-color cloud studies based on the view out of the aircraft window. Two years later on a flight during which there were no clouds to picture, she started working with the landscape below. Her first major aerial painting was titled *Passagassawakeag I*, 1975, a view over a place near the family summer home at Searsmont, Maine. She chartered a Cessna, flew several times over and around the town at about one thousand feet, walked around the area, and even made a model of it so she could better understand the topography. The next year she was painting from as close to the top of the WTC as she could get, and then returned as a resident artist in 1997, making pastel studies from offices in both towers. She painted *Flatiron Intersection* in 1979, a view from the 15th floor that included a corner of Madison Square Park, and the next year worked from the Empire State Building. She ended up working over Chicago, Minneapolis, San Francisco, Tokyo, and Hong Kong, sometimes using views from aircraft, sometimes from hotel windows. In 1998 she worked from the actual WTC observation deck to make a nocturnal view, and then in 1999 from the 107th floor to paint *Vertiginous: World Financial Area*. She worked from various windows and angles, making a painting that makes the viewer feel as if one is telescoping downward, yet seemingly intimate with the scene below at the same time. You can't help but think of Michael Richards when you look at the painting, of the people in the building leaning out of the windows and making the decision whether or not to become airborne.

# 11. Ground Truth

*Mechanical wings allow us to fly, but it is with our minds that we make the sky ours. . . . Flying at its best is a way of thinking.*

—William Langeweische

**Flying in Joe Thompson's plane down the Hudson River Valley and** around Manhattan and contemplating the aerial art history of Manhattan gives me only a few of the viewpoints in the constellation I want to assemble. We had finished the flight that day by swinging east around the tip of the city and flying the length of Long Island with its mansions in East Hampton out to its end at the lighthouse in Montauk. We set down near the latter for a seafood lunch, then headed back north, following the Connecticut River before breaking back west by Mount Greylock and into North Adams. A few days later I retraced our path down the Hudson by car with Matt Coolidge, one of the founders of the Center for Land Use Interpretation and for all intents and purposes its managing director.

The first time I actually stopped in Wendover for more than a refueling while driving east from Reno to Salt Lake City was in 2001. Matt and I were taking a driving tour of America's playas, those intermittently wet dry lake beds, as material for a book I was writing about what we do on them. Those activities range from blowing up atomic bombs to creating artworks, and CLUI had assembled a library and database about most of them. Matt is an affable and casual road companion until you hit the button that says, "retrieve data." At that point he starts reciting reams of history and statistics interleaved with insights about how we use, reuse, abuse and misconstrue landscape that can leave you gasping for air. Or, in my case, with hand cramps from trying to take notes fast enough. Never in my life have I longed more for a recording device than when traveling with the man.

CLUI shares half a building in Los Angeles, more properly Culver City, with another slightly inscrutable organization named the Museum of Jurassic Technology. "The Center," as Matt prefers to call his own group, has established several field sites around the country, the first of which was in 1996 at the Air Base in Wendover. Its most recent expansion is the acquisition of a large building in downtown Troy, New York, just over the Taconic Mountains from Williamstown. Matt means to do for the Hudson River Valley and environs what the Center has done in the Mojave and Great Basin deserts, which is to catalog and present the varieties of land uses to its residents. When Matt drives up from Troy to collect me, he's piloting one of the Center's two white Jeep Cherokees, vehicles of indeterminate age and mileage that crisscross the country carrying maps, GPS units, cameras, building supplies, remnants of exhibitions, and this late afternoon a green canoe. As we drive over the mountains, he explains that, after riding around in a helicopter in an attempt to locate an earthwork by Nancy Holt atop a landfill in New Jersey, we're going to use the canoe to paddle up the Hackensack River to a chlorine plant. Oh, I say. And then he launched into an unexpected bit of personal aerial history. Here's his mostly unedited account—reconstructed via my notes and his e-mails with all their glorious asides, because I didn't have a recorder and that's how he talks—of why he had the canoe atop the Jeep.


"One of my great-grandfathers (we all have at least four of them), Harold Jefferson Coolidge, bought an island in the middle of Squam Lake, New Hampshire in 1892 (for one hundred dollars, which was given to him as either a twenty-first birthday or graduation present by his aunt ((Isabella Stuart Gardner, who had an interesting museum of her own in Boston, which is still there (((and where a kid I knew in college was working as a security guard one night when he answered the door to let in some people dressed in Boston Police uniforms . . . but that's another story)))).

"His son, my great uncle Harold Coolidge, was a naturalist and conservationist who pretty much ran things out on the island when I was a kid (my grandfather, Harold Jr.'s brother, died in the 1950s). We would spend a lot of time at 'Uncle Hal's' place, as he would often have all kinds of interesting people there. In the evenings he would show movies of his expeditions into Africa and Indochina, collecting specimens, filming gibbons, and such. One of his friends was Brad Washburn, who also had/has a house on this lake. Brad would often fly to the lake and land on it in his seaplane, driving the plane into the cove at Uncle Hal's (formerly my great-grandfather's) house, and he would take the children up for rides. If I went up in the plane I must have been very young, as though I remember the plane, I don't remember going up in it.

"A third piece of ground that Washburn mapped with unprecedented detail and accuracy, besides Mount McKinley and Everest, was Squam Lake, New Hampshire. The map he made of it is still the map everyone uses of the lake, and bears his name. The lake has many islands and coves and channels and mountains around it, and is quite a complex landscape. Over the years (mostly in the 1950s and 1960s) he surveyed the lake with precision. Members of our family helped him with some of this, especially taking soundings of the water depth. Brad mapped just about every rock on the lake. He left surveying posts and marks on some of the shoreline rocks, which became mysterious icons to me. Land marks.

"In addition to being the director of the Boston Museum of Science, which I knew well as a child, Brad also established the nearby Squam Lakes Science Center, an outdoor nature learning center in Holderness, New Hampshire, with a walking trail and exhibits which, now that I think of it, must've made quite an impression on me too. It certainly was the first walking trail that was ingrained in my mind, and it was rather complex and lengthy, and had raccoons, snapping turtles, a blacksmith shop, bears, stick-your-hand-in-the-hole-and-guess-the-type-of-fur type exhibits. There were also science lectures held in the barn there that we went to. The science center is still there, and now has a very nice exhibit about surveying and Brad Washburn, an exhibit that Washburn helped create.

"Great-grandfather Harold Jefferson Coolidge died as an old man on the island. I've known this place all my life, and I'm now building a cabin there. It was from there that I was driving when I came to pick you up."

Stunned, I was. I had known that Matt had wanted to be a pilot when younger, and had actually taken lessons up to the point of soloing, the last step before you get your license, when he quit. He had realized that it wasn't the flying that he enjoyed, but the looking and taking pictures, and he couldn't do both. He chose taking pictures, and to this day the CLUI archives include aerial photographs, many taken by him, of mostly cultural features in the database. But I had no idea about Bradford Washburn's role in Matt's life, a man whom I considered to be a twentieth-century embodiment of Alexander von Humboldt, and a personal hero.

Washburn was born in Cambridge, Massachusetts in 1910 and suffered from chronic hay fever. At the age of eleven, hiking up Mount Washington, he found he could suddenly breathe much more easily. Climbing became a passion, as did photography at age thirteen when his mother, an amateur photographer, gave him a Kodak Brownie. By the time he was sixteen he had already published a local climbing guide to the White Mountains and climbed the Matterhorn and Mont Blanc in Europe. While he attended Harvard, earning an undergraduate degree in 1933, he supported himself off the sales of three books he had written about his climbing exploits and illustrated with his own photographs. The same year he graduated, he began a chain of seven first ascents on major peaks in North America. He was the first person to twice climb Mount McKinley (otherwise known to climbers, not to mention native Alaskans, as Denali), and his wife Barbara became the first woman to stand on its summit.

The young climber and author taught at Harvard's Institute of Geographical Exploration from 1935 until 1942, and in 1936 also became the director of what was then a dusty and amateurish collection known as the New England Museum of Natural History. By the time he retired as its director in 1980 he had turned it into the Boston Museum of Science, the first museum to deal with all the sciences, be they natural, physical, or applied—a reflection of his own career. In the midst of making a name for himself as one of the preeminent mountaineers of the twentieth century while teaching and administering, he inherited a fifty-three-pound Fairchild aerial camera from Captain Albert William Stevens, a pioneering balloonist who in 1930 took the first photograph showing the curvature of the Earth. Stevens had taught Washburn how to use the large-format roll camera and he used it, along with another he had bought previously and rebuilt, to make what remain among the most detailed portraits of mountains and glaciers in Alaska and the Yukon. Climbers today still scrutinize the black-and-white images before attempting to negotiate the formidable defenses of those summits, yet they are also exhibited in art galleries and museums. Ansel Adam referred to Washburn as a "roving genius of mind and mountains."

In 1960 he earned a dual master's degree in geography and geology, and then proceeded to make definitive maps of Denali, the Grand Canyon, Mount Everest (Chomolungma), and the Presidential Range of the White Mountains, his home range. The maps, among the two dozen or so major ones he compiled from firsthand fieldwork, are models of precision and clarity, and the Everest map, which was based on the first-ever allowed flights over Nepal and Tibet in 1981–1984, was published by the National Geographic Society to enduring acclaim. Not content to climb, write about, and photograph mountains, he also studied them as benchmarks in continental drift. In 1992 he succeeded in bouncing a laser beam off the summit of Everest and corrected its elevation upwards by seven feet, a technique he had perfected while mapping the Grand Canyon in the 1970s.

As a curious side note, in 1937 Washburn was a candidate to serve as Amelia Earhart's navigator on her legendary attempt to fly around the world. When interviewed by her, Washburn, ever the careful and experienced expeditioneer, suggested that they place a radio on Howland Island, an uninhabited atoll halfway between Hawaii and Australia. He thought it would serve as a beacon along their route over the middle of the emptiest, hence most dangerous portion of the journey. Earhart, a supremely self-confident aviator, disliked the advice and instead selected the well-known transpacific pilot Fred Noonan. Later that year they disappeared in the vicinity of the island.

I've always considered Washburn a hero because he was able to climb high mountains in extreme and difficult environments, and while doing so conduct scientific and cartographic work of the highest caliber, and then to put that work into a larger context for the public through collective efforts of the science museum. Humboldt would have been thrilled to see his principles applied on such a public scale.

Likewise, what has continually amazed me about Matt Coolidge, and the work done by his colleagues at CLUI, including Sarah Simons, Steve Rowell, and Erik Knutzen, is that they have used those tools—elevated vantage points, masses of comparative images and data, transdisciplinary analysis—not to study the natural world (if such still exists), but the most overt anthropogenic landscapes. CLUI publishes guidebooks to places such as the Nevada Test Site and the military industrial facilities of the Great Salt Lake Desert, but also puts up exhibits about fiber optic networks, and drowned towns. What Matt and I will be doing the next day is compiling images and information to be used in the Center's "Up River" exhibition in Troy, a tour of the manmade sites along the Hudson from Manhattan's Battery Park north. The exhibition will demonstrate how the landscape of the Hudson has for decades been moving from a landscape of resource extraction to one of cultural tourism. It's not just the establishment of contemporary art collections in recycled industrial buildings, and museums for the likes of Frederic Church and Norman Rockwell, it's actual changes in land usage that signify a shift in the sociopolitical ecology.

The next day, having spent the night in Troy, Matt wakes me up at 5:45 A.M. with a cup of foul coffee, but it's a nice morning out, and cool enough for a sweater, rare in August, much less during a summer that's been unusually hot and muggy. We squeeze in underneath the canoe and drive through the mostly depressed downtown. Matt starts his running commentary with a disquisition about why Troy, which used to be one of the most prosperous cities in America with a population of almost seventy-seven thousand at the beginning of the twentieth century, has now a hundred years later slipped below fifty thousand and is one of the less flourishing cities in the state.

"Bessemer invented his steel process in Troy—you blow air through molten iron to remove impurities (so cannons don't blow up, among other things). Troy ended up making railroad spikes, thus 'hammering the nails into its own coffin' as the railroads made the canals obsolete. And then Bessemer gave away the process to Pittsburgh. Amazing . . .

"Troy's half empty," he observes. "The twentieth century just skipped it. There's almost no tax base here, the Rensselaer Polytechnic Institute taking up most of downtown as a tax-free nonprofit, so what businesses are here are overtaxed. And the city turned its back on the river." It's true. We go by the bank of the Hudson and are separated from it by a chainlink fence. Across the street is a city parking garage. Elsewhere it would be all restaurants and shops facing the waterfront. We cross the river toward I-87, and pass the Watervliet Arsenal that I'd seen from the air. I recalibrate my sense of how large the gun barrels are from my view on the ground. They're much, much larger than I thought.

At seven o'clock we cross the Catskill Creek defile, Thomas Cole having made his famous view of it just east of our route. I-87 is the quick route down to New Jersey, and we're on the west side of the mountains from the Hudson, but as we draw abreast of where Dia:Beacon sits on the east shore of the river, we're also coming up on the Storm King Art Center, at five hundred acres one of the larger and older contemporary outdoor sculpture parks in the world. When its owners bought the farmland in 1960, they originally thought to establish a museum for Hudson River painters; but, after, one of them visited a marble quarry in Austria, they changed their minds and began to focus on sculpture. They started with a formal garden, then in 1966 and shortly after the death of David Smith, they bought several of his steel works for the property. After that Mark di Suvero became a focus, his enormous and carefully engineered steel-beam-and-cable works for year a defining focus of the collection. I like di Suvero's work; along with Richard Serra's massive ellipses torqued out of thick steel plate at a shipyard, they pretty much define the upper limit of macho American metal sculpture.

Andy Goldsworthy is the celebrated Scottish artist who seems to walk around somewhat aimlessly, picking up whatever comes to hand—sticks, leaves, flowers, snowballs—and arranging stuff into intimate and somewhat contradictory gestures, like Zen koans. He photographs his interventions at close range; the ones made from stones, which often create the illusion of deep spaces enclosed, disappear from a few feet away. But at Storm King he built a wall 5 feet high by 2,278 feet long. It starts along the line of an old stone fence, then weaves downhill in a serpentine fashion around individual tree trunks. The wall declines gently to the edge of a pond where it seems to disappear underwater, only to emerge from the other side, climb over and down a hill and then heads straight to the western property line in a field that abuts the freeway we're driving down.

This is not a small gesture, this wall. The German filmmaker Thomas Riedelsheimer, who made the best-selling documentary about Goldsworthy, *Rivers and Tides*, spent most of his time filming the artist in close-ups and even on his hands and knees. He starts out close in on the wall with the hands of Scottish stonemasons carefully dry-stacking the rocks, but soon has to pull back in a helicopter in order to follow the piece to its end. It's too large to capture from the ground, its own landscape feature, an anti-quarry, and the film reminds me of nothing so much as Smithson's helicopter views when *Spiral Jetty* was completed.

Earlier this summer I walked the wall from one side of the pond, through the forest and up to its starting point—from which the di Suvero sculpture is mostly out of sight—and found that when I walked back among the steel pieces I was far more at ease with them. Other works by Goldsworthy, and much of di Suvero often appear to be or actually are precarious, and can or do fall down. That's part of their elegance and power: they are made of earthbound materials at least potentially on the verge of collapse.

Storm King Mountain has an aesthetic history beyond the sculpture park, and one that is a hallmark in the conversion of the Hudson River Valley from resource extraction to cultural tourism. At the beginning of the last century, Edward and Mary Averell Harriman owned thirty thousand acres along the river and had a splendid retreat across the river from the timbered slopes of the mountain, which stands at the northern end of the Palisades, those basalt cliffs rising from the valley. Quarrying and logging began to take a toll on the scenery across the river, but when the town of Sing Sing came up with a plan to build a new prison at Bear Mountain, Harriman, president of the Union Pacific Railroad, led the fight to preserve his view and rallied his friends J. Pierpont Morgan and John D. Rockefeller to the cause. He donated ten thousand acres to the state and $1 million to see to it that the Palisades parklands were extended north. The Harriman-Bear Mountain State Park was soon a success. Founded in 1910, within four years it was seeing a million visitors per year.

In 1962 a major contemporary conservation battle to preserve the river was joined not by millionaires, who had long since moved on to other and less crowded pastures, but by a coalition of environmental groups. Consolidated Edison proposed to build a hydroelectric plant with a reservoir bay cutting deeply into Storm King just north of Bear Mountain, and running transmission lines through the viewshed. It took seventeen years, but the power company backed down. Where Bear Mountain sees more visitors per year than Yellowstone, according to the park service, the newer Storm King State Park has been left undeveloped for the hikers and hunters needing more solitude.

Just before 8:00 A.M. and the New Jersey stateline, we bear east to gain the Palisades Interstate Parkway. John D. Rockefeller donated land to keep this road a pleasant one, and worked out a trade: road construction for the tearing down of houses in sight of the river. Unlike the interstate system, the Cold War specifications of which are designed around the ability to carry heavy tanks, the parkway is just that: it follows the contours of the ridge, is planted with native flora, and is an entirely pleasant way to enter the Garden State.

A sign announces that we're about to intersect I-80 on our right. "Next stop, Wendover," Matt calls out as we cross the interstate, which several thousand miles away enters Nevada at the desert town. "I-80 is America's Main Street, going from one empire city, New York, to the other, San Francisco." To our left are the feeders heading to the Washington Bridge, its towers visible as we segue over toward the New Jersey Turnpike and the north end of the Meadowlands, 8,400 acres of wetlands orphaned between freeways and railroad lines. This is the open area remaining from what in the nineteenth century was an estuarial marsh of more than 20,000 acres, or thirty-two square miles. According to the Meadowlands Conservation Trust, in 1995 there were 2,500 acres devoted solely to dumping, fifty-one individual operations receiving eleven thousand tons per day of northern New Jersey's garbage.

In front of us is the steel tracery of the Pulaski Skyway, which will be our constant landmark throughout the day. "An Erector Set wet dream," Matt says with a grin. The Newark airport is on our right, the Port of Elizabeth on the left. It's what's left of the old Port of New York operations, and is the largest container port in the Northeast, but small compared to the Los Angeles and Long Beach complex. Trade with Europe, given its relatively high manufacturing wages and subsequent cost of goods, is nothing compared to the volume of trade with Asia where the opposite is true. We're surrounded by refineries, landfills, warehouses. This complicated landscape so utterly anthropogenic is where we'll be spending our day in the air and on the ground, and it is as far from a garden as I can imagine, by far and away the most toxic urban environment Matt and I have ever traversed. The only landscape I can think that's worse is the Nevada Test Site.

It's not literal, but I feel as if someone had taken a broom and swept out the garbage from the Hudson and here's where it ended up, downstream and onshore. To call the Meadowlands a wetland area is to give the false impression that it's a bit of nature preserved. It's not. It was hunted, harvested and fished out a century ago, its waters diverted and dredged, the land reconfigured and used as new real estate. The dumps hold not just regional toxic trash, but even bits and pieces of London masonry, rubble from the World War II bombing of Britain that was then loaded into ships for ballast, and dumped here afterwards. It's astonishing, given the toxicity of the location, to see flags flying at the sales offices of real estate developments—and then I remember the same conjunction in Los Angeles near the ports. Alan Berger, who teaches landscape architecture at Harvard, has been looking at such reclamations, spurred on by his realization that as he flew in and out of airports, he saw the same pattern time and again.

Berger grew up in Connecticut, earned a bachelor's degree in agriculture from the University of Nebraska, and then a master's at the University of Pennsylvania in landscape architecture, a discipline in which he has been licensed since 1992. This is a man who is used to having his hands in dirt. Given a teaching job in Denver, shortly after his arrival in 1996 he drove into the Rockies late one day and set up camp in the dark. When he woke the next morning he found he was next to the alienscape of an open pit gold mine. He'd never seen anything like it before. The colors of such places tend to run to surreal oranges and reds that clearly don't belong on the surface of the planet, and the color of the cyanide-tainted waters leaching from them would be beyond belief if you haven't seen them for yourself (or in photographs by people such as Alex MacLean).

The young architect began flying with local and federal government agencies in order to document some of the more than two hundred thousand abandoned mines spread out over one hundred million acres in the West. It took him more than five years to complete his book *Reclaiming the American West*, an aerial and cartographic exploration that documented sites, proposed analytical tools, and set the stage for his ongoing work on reclamation proposals. As Berger was flying in and out of urban airports, he noticed the wastelands left behind as urban growth moved outward. These "drosscapes" then become a kind of new frontier for redevelopment as the need for land increases around the urban core. The analytical mapping techniques that he was developing over the mine tailings of the rural West, he realized, could be applied to these byproducts of industrial sprawl. For Berger, the drosscape was as inevitable a result of industry (and defense facilities) as tailings were for a mine, and he's published a book about them, *Drosscape: Wasting Land in Urban America*.

As Matt and I drive towards Teterboro, the general aviation airport we'll be flying out of, I'm turning the road map in circles trying to figure out where our primary targets are, the huge Fresh Kills landfill and a smaller one where Nancy Holt was commissioned in 1988 to make *Sky Mound*. Matt also wants to take a look at some sunken ships in the harbor and the local power plants.

"Road maps are useless for getting around in the Meadowlands, because the roads are for getting you through it." Or, in the case of the Pulaski, over it! "What's best are topo maps, aerial photos, and GPS." When we arrive at Pegasus Global Helicopter Corporation a few minutes later, it's seventy-nine degrees out, muggy, and despite a decent breeze very hazy. He hands our pilot, Larry King, an aerial photo of the landfill, where in theory we'll find Nancy Holt's *Sky Mound* project. King thinks he recognizes the roads, but notes that the new post-9/11 security rules say we can't photograph the power plants and have to stay away from them.

King gets on the phone and calls Teterboro air control to let them know we're doing a photo shoot, although a reconnaissance would be more accurate, if a poor choice of words. They want a letterhead faxed over with all the details, the first time we've seen such a level of verification for a general aviation flight. But we are next to Newark Liberty International Airport, which last year handled thirty-three million passengers. Along with Kennedy and La Guardia, the air traffic about New York City makes it the most crowded airspace in the country. A little paranoia is not necessarily a bad thing.

King walks us out to the R-44 we're flying in. All the major television stations in the area have their helicopters here, which are in various stages of readiness in between the morning traffic reports and noon news cycles. We taxi just above the ground down to the runway, and take off right above a flock of geese. I'm anxious about it, but they don't even fluff a wing at our departure. New Jersey sangfroid, I presume. We traverse a refinery at about seven hundred feet and head for the Hackensack River and Staten Island where Matt begins happily shooting the crumbling jetties and rusting hulls of tugboats and barges listing in the water, rust buckets left to their fate.

King, negotiating all the while with Teterboro, angles us over the Fresh Kills Landfill. The name is both misleading and revealing at the same time. A "kill" is an affix from the Dutch that means a stream. Around here that means a waterway that's been murdered. Fresh kill refers pretty directly to the contents of the landfill, which is to say mostly household garbage from across the river. Fresh Kills was first opened as a temporary dump in 1947, but stayed open until a court order forced its closure in March 2001. It was the largest landfill in the world, covering 2,200 acres and at 225 feet in elevation, taller than the Statue of Liberty. It is, in fact, the largest hill on the Northwest coastline. In 1995 the EPA estimated that it was emitting 2 percent of the world's entire load of methane, and because it was unlined, leaching four million gallons of liquid into the Hudson per day. It's so large that Matt has King take us up to 1,200 feet so he can get the largest of its mounds into the viewfinder.

Across the river there is a hole where the Twin Towers stood. Fresh Kills was reopened almost immediately after 9/11, the nearest facility that could accept the amount of rubble from the destroyed buildings. It also served as the work station where remains of bodies were identified, including those of artist Michael Richards. Much of the rubble was later transferred for recycling, and today dump trucks are rotating around the lower slopes of the mound spreading out what looks like asphalt. Plans call for thirty years of settling, resculpting, methane harvesting, and then creation of hiking and biking paths, a wind farm, and other recreational amenities.

Matt keeps trying to get us closer to the small landfill where we think Nancy Holt was supposed to build her earthwork, a thirty-seven-acre site unfortunately almost underneath the approach pattern being used today by Newark air traffic. There are so many closed landfills in the area, it's hard to make sure, but by comparing the configuration of the roads and freeways from map to ground, we're pretty convinced that it's the logical candidate. King keeps talking with Teterboro, but the young controller, obviously a bit frantic, keeps pushing us south and east. We can see the traffic landing, the low approach pattern, so we understand the concern. One small aircraft triggering the collision avoidance alarm in a jetliner would screw up the approach pattern for an entire day. I look beneath us to see what environmental horror we're over, and to my surprise see it's a golf course. We hover over the course while trying to talk our way over the landfill, which must be driving the golfers nuts.

All around and framing the golf course are Berger's drosscapes. The landfills awaiting conversion into baseball diamonds, storage sheds generating a little income until condominiums can be built, corridors for power lines, contaminated sites left unreclaimed from military-industrial occupation. I could swear Berger developed his list flying over this exact terrain, instead of Denver, but that proves his point: it's a ubiquitous pattern of post-Fordist production and late capitalism. Berger also notes that the country's solid waste doubled in the thirty years since 1960, but that landfill sites decreased from around 20,000 in the 1970s to only 2,268 in

1999. The consolidation is due, in part, from environmental regulations that make it more cost-effective to build large sites far outside cities, versus smaller ones inside the boundaries. I wonder where the new and even larger landfills are for New Jersey.

King has given up trying to get us anywhere near the Holt site, so we cross the river over to Manhattan to try another vector through the Newark control tower. Two other helicopters are nearby scouting locations for filming, aerial shots of Manhattan as common in the media as those of downtown Los Angeles. By 10:45 we're back on the ground having been totally rebuffed by air control. Now it's time to see what we can accomplish on the ground, and we start, of course, by trying to get as high above it as possible by driving over to the Pulaski Skyway. As we're leaving the airport, a small but powerful banner plane skims low over the runway to snag a wire suspended between poles. It misses the first time, loops around, and hooks it on the second run. The banner rises off the ground and straightens out in the air, an airborne billboard for the B-thriller *Snakes on a Plane*. I shake my head, unable to imagine making up aerial material as good as what the world simply provides daily.

At 12:30 we're on the Pulaski, a roadway that opened its two cantilever truss bridges over the Passaic and Hackensack rivers in 1932. The elevated skyway carries four lanes of traffic between Newark and Jersey City, and was designed to relieve congestion in the port area as goods were transferred from ships to rail lines and trucks. Unfortunately, what was proudly billed as the world's longest elevated viaduct turned out to be too narrow and unsafe for trucks, so it has almost always been used solely by cars. Not, mind you, that it's all that safe for any vehicle. The elegant but tight passage over the rivers, marshes, and industrial areas sees about sixty thousand vehicles every day, most of which seem to be traveling well over the speed limit. After more than seventy years of use, plans are being made to replace the historical structure with a strengthened replica in order account for contemporary traffic and speeds, and to finally allow trucks to use it. It's going to cost more than a billion dollars to do so, and that's before the cost overruns start. And there will be overages.

Since 9/11 you're not allowed to photograph the Pulaski unless you're an approved journalist, so I confine myself to taking notes about the way the flicker below appears to be a clear view in between the steel struts, which are proportioned just right to allow that at 55 mph. Of course, you'll be run over if you try to maintain that speed, but it's nice to note. That view is, in part, why we're taking the skyway, hoping to locate which road will take us over to the landfill that we can see in the distance.

We stop for a late lunch at a diner underneath the west end of the skyway, the kind of place where you order a hot turkey sandwich on white bread smothered with gravy. Outside, ferrying steel shipping containers, it's a constant stream of semis, many of which have been customized with cannibalized sections to increase their carrying capacity. It's like eating lunch in the middle of a trucking yard, the smell of diesel as thick as butter on the bread.

Back in the Jeep, Matt plugs a GPS unit into his laptop, which is running a USGS topo map program, so we will have an aerial schematic of where we're going with our presence noted as a moving cursor on the map. Blaut and his kindergarten kids would have loved it. We circle and circle the location, closing in until we've managed to drive twice around the four-sided polygon of roads, ditches, and reeds that enclose what turns out to be the correct landfill. It's also and somewhat surprisingly to me the one we were scouting in the air from afar. Matt ignores the NO TRES-PASSING signs at several businesses and drives up behind their buildings to the fence that actually encloses the landfill. We consider assaulting the barrier, but let it go as we have other plans for later in the afternoon. It was obvious from the air and even more so here that nothing Nancy Holt planned was ever implemented.

The closure of this landfill in the 1980s was a model of environmental propriety compared to practices in the past. A slurry wall was built to stop the leachates from migrating into the groundwater, and recycled plastic liners were placed over the top and capped with soil. That was billed as stage one of Holt's project, but in actuality the $11 million spent was required by law. It was the second part, the actual art, that was never funded and built. In Holt's 1985 drawings for *Sky Mound*—which would have made a nice allusion to the nearby Pulaski skyway—gravel paths radiate inward past eight "star-viewing mounds" to meet at a central point. The paths and a series of steel poles would have marked the equinoxes and solstices, and the methane wellheads the position of the moon.

Holt told Anthony Aveni, the Nazca researcher, that the reason she wanted to do the project was to re-create the feeling she got standing atop the solsticially-aligned Monk's Mound in Illinois, and other astronomically sited ancient structures. As with so many earthworks new and ancient, her inspiration was that old desire to connect sky and ground, Heaven and Earth. She also stated that Walter De Maria's parallel lines in the Mojave and Smithson's works were inspired at least in part by the Nazca lines. But where she and those other artists were able to complete their projects in the desert West, the convoluted politics of New Jersey defeated her here.

Construction of her earthwork would have required state and federal arts grants, as well as becoming part of an environmental bond issue, all of which was declared moot when local officials announced that the rate of settlement inside the landfill was too great to consider building anything atop it. The project was abandoned. The comparison of *Sky Mound* with Holt's earlier *Sun Tunnels* in Utah points out why large land arts projects are found more often in the deserts of the West than elsewhere. In historical terms the deserts were considered a different kind of wasteland than the landfills of the East Coast. The land was, so to speak, dirt cheap. Garbage dumps turn out to be expensive real estate because of their proximity to the residential neighborhoods that spring up around them as open land becomes increasingly scarce. Expensive because they have to be remediated.

Coolidge and I break off our obsessive circling of the landfill and head across the Hackensack River for Laurel Hill Park, a different kind of aesthetic rescue mission. We end up at a boat launch ramp and untie the canoe from the top of the Jeep, carry it down to the water's edge, and prepare to paddle downriver. Two guys in wetsuits riding Jet-skis roar up, and look at us askance. "You're kidding," one says. "Nope," replies Matt. The two guys shake their heads as they back up their SUV and trailer to the ramp.

Laurel Hill used to be, well, a 150-foot-high hill, an old plug of Triassic volcanic rock. If you've ever seen the ads for Prudential Insurance, which bills itself as being as solid as a rock, that's not a profile of Gibraltar that inspired the ads, but Laurel Hill, which an advertising executive in 1896 (yes, they existed even then) thought would make a strong logo. During the 1960s Hudson County leased the property to a quarrying operation that dynamited and carted away three-quarters of it, lowering the summit by fifty feet. The blasting was stopped because the vibrations from the explosions came uncomfortably close to the New Jersey Turnpike, under which we will be paddling on our way to inspect the chlorine plant on the other side of the river. The hill is as much an industrial ruin as the concrete sides of the plant dimly visible to our south.

We launch out into a stiff breeze. There's some current going our way, but it's a battle for us among wind, tide, and crosscurrents. Still, it's great to be out on the water, even though it smells awful, feels oily on the skin, isn't safe to swim in, dead crabs are floating belly-up beside us, and a muskrat I spot on the shore looks like it's been bleached in acid, its fur an astonishingly unhealthy hue. And this area is what's billed as the gem of the Meadowlands reclamation. That shows you how far back from ruination the area has come. The park was established in only 1997, part of a larger effort to clean up the Meadowlands, and in the last decade the Saw Mill Creek Marsh area—the feathered reeds of which are on both sides of the river—has become one of the better habitats in this part of the riparian region.

Hawks, egrets, heron, and osprey hunt and nest nearby. Diamondback terrapin and snapping turtles, opossum, raccoon, and even red foxes are living here. You wouldn't want to eat the crabs or fish, but maybe someday—a very long someday from now—you might be able to. I am, it should be noted, an incurable optimist, whether it's surviving helicopter flights or canoe rides. Matt and I keep paddling. Despite how much fun we're having, it's clear that we've picked a less than efficient mode of transportation for today. Jet-skis would definitely be the way to tackle this project, so we turn back and enjoy a faster ride back to the landing.

We float briskly under the New Jersey Turnpike and the Hackensack River Bridge, invisible to the drivers above us a bit more than fifty feet away. The bridge, built in 1951, is 5,623 feet long and was designed to afford drivers a great view ahead of them of the Manhattan skyline, but with retaining walls high enough that they would not be distracted by the river and marsh below. The aerial view was blocked off as a hazard. Robert Smithson actually made an artwork on Laurel Hill in 1968 that reflected how we displace views. He propped up black-and-white photographs of the surroundings as if they were mirrors in the landscape, and then photographed the setups as *Untitled (6 Stops on a Section)*. The work was all about what you could and could not know in landscape, the presence and non-presence of a witness. One of the photos was set flat on the reddish rock, a black-and-white picture taken looking straight up into the sky and the sun, a blank white eye, no one visible looking down. If you stood over the photo, there would be no reflection. You would be absent, suspended invisibly between the heavens and the earth.

# FLYING NORTH
# TO SOUTH

*From the Plane Window*
*At thirty-nine thousand feet*
*I looked down*
*on the backs of a million sheep*
*In pastures azure blue*
*On the horizon were minarets*
*and spires of castles*
*rearing high*
*I caught a glimpse of land*
*my land*
*far below*
*Some call it desert*
*but it is full of life*
*pulsating life*
*if one knows where to find it*
*in the land I love*

—Jack Davis (1917–2000)

**Most of us English-speaking folk spend our time in the air flying** parallel with latitudes in the northern hemisphere, which is to say east-to-west or west-to-east, less a touristic predilection than an historical bias created by the flow of capital between Europe and North America, and, since 1854 with the opening of Japan, increasingly with Asia. That began to change during the latter part of the last century as the flow of goods and money, ideas and people, became more transequatorial, but the number of passengers flying horizontally around our planet remains much greater than those making vertical transits.

That's why I always feel that to board a flight heading south and east across the Pacific Ocean from Los Angeles to Sydney is a significant departure. Crossing the world's largest ocean takes most of the fourteen and a half hours along the route, and at thirty-eight thousand feet during the wee hours of the morning with a full moon outside, the thunderheads of the tropics glowing silver like nature's own version of the apocalypse, it's positively ethereal. Given that I don't think widespread commercial air travel is a particularly sustainable activity (think peak oil, global warming, inflation), I treasure our current ability to see such a large arc of the planet from the air.

You would think I'd want a window seat on such a flight, but I stick stubbornly to the aisles so I can roam around the plane. And the crew lets me hang out at the window back by the rear galley, so all I really miss are the takeoffs and landings. Tonight, when I should be asleep after our midnight departure, I'm standing near that aft galley, bent over at the waist, face to Plexiglas. The double panes of plastic are small and the angle of view less than ideal, but from here the ocean is a dark blue slate glimpsed between the clouds, the world seen almost as if in negative. Darker masses that might be small islands appear and disappear.

I'm headed back to the capital of Australia, Canberra, where I'll be based at the Australian National University while attempting to understand why most Aboriginal paintings are aerial views. My aircraft this time is a Boeing 747-400, a late-model version of the wide-body double decker that first started carrying passengers in 1970. I can remember the first time I saw one, how astonished I was at its size next to a 727, which up to that point was the largest aircraft in which I'd flown. The fact that the 747 had a second deck up front, producing the signature bulge above and behind the cockpit, was scandalous. How could such a thing fly? As of today more than 1,300 of them have been built.

Boeing had originally thought it would make only four hundred of the large passenger jets, which were supposed to be replaced by supersonic aircraft. Didn't happen. The only supersonic passenger jet put into service that was even barely viable was the Anglo-French Concorde. Twenty of the craft were built and they started commercial service in 1976, able to cross the Atlantic in about half the time of a regular jet. When they were pulled from service in 2003, twelve of them were still flying. Although the Concorde was disliked by environmentalists and others for its high noise levels, sonic booms, and increased level of havoc wreaked on the ozone layer, what killed it was the 747, which carries four times the number of passengers for the same amount of fuel. In our case that would be more than five hundred passengers using around fifty-seven thousand gallons of fuel to fly seventy-five hundred miles. That's a large carbon footprint gobbling up increasingly expensive fuel. Despite the airlines building ever larger and hence more efficient-per-passenger airplanes, I suspect it won't take long for such trips to be the sole province of the rich and the weaponized.

I actually like 747s because I consider them safe. Despite the loss of forty-five hulls in various accidents, and with 3,707 souls lost in said occurrences, it's a massively overbuilt machine with redundancy everywhere. The planes are 232 feet long and contain six million parts, half of which are rivets, as Barry Lopez notes in his essay "Flight." The redundancy consists of parallel systems such as four separate hydraulic ones and four main landing gear sets for a total of sixteen tires. And as I observed over Utah, its glide ratio of 15:1 means that at our current altitude we could cruise for almost one hundred miles before landing. Not, mind you, that ditching in the middle of the ocean would be a picnic, but at least we'd have time to let people know where we were.

And where we are right now depends on your viewpoint. It's March 31st at home, April 1st in Sydney, and we're about to cross the International Date Line. No matter how many times I've done it, by sea or air, I still have trouble understanding how it works. So did Magellan's crew when in 1522, after sailing westward for three years during their circumnavigation, they landed in the Cape Verde Islands thinking it was Thursday when it was only Wednesday. Sir Francis Drake and other explorers of that century experienced the same anomaly, and by 1612 the first proposals for establishing a date line were being floated about. In 1830 a German astronomer proposed that it should be a meridian opposite the prime meridian located at the longitude of the Royal Observatory at Greenwich outside London. That British geographical determination was the baseline from which Her Majesty's mariners measured their time so they could determine their distance from it, their longitude, as they sailed about the world, an absolute necessity for accurate navigation between far points out of sight of land.

From the utility of the prime meridian evolved a standardized set of time zones based on Greenwich Mean Time (GMT), which came about because railroads began ferrying passengers and cargo with increasing frequency across distances large enough that no one could keep track of themselves or their goods any longer. Before widespread acceptance of the GMT, every town set its own time; needless to say, the potential for a literal train wreck was unacceptable. Airplanes, because they move faster, required an even more precise set of navigational standards, and as their speed increased so did the accuracy of the clocks involved. In 1972 the GMT was superseded by a Coordinated Universal Time based on atomic clocks.

To add to my confusion, the date line itself keeps changing. As it's not a physical line, or even one spelled out in international law, it zigs west to thread the Aleutian Islands in the Bering Strait and to accommodate the border between Russia and the United States. We're crossing it near the Samoan Islands, where it is Saturday. Down here the line zags east by hundreds of miles to bring the Tongan Islands—just to our south—into the same date as New Zealand, which administers them. There it is Sunday. And yes, in colloquial English we conflate place and time: "Where it is Sunday."

In 1994 the tiny republic of Kirbati decided to unify its thirty-three atolls and eighty-five thousand inhabitants, which are spread out over 1.35 million square miles of water and precious little land, within one date. Its president simply declared the date line nearby moved more than two thousand miles east. In theory that allowed all the businesses within its borders to synchronize their activities, thus giving them a regular five-day work week. It also allowed the islanders to claim the privilege of living on the only place on earth outside the Antarctic to receive the first dawn of the new millennium. The mapmakers of the world continue to ignore the Kirbati declaration, however, and my atlases do not show it. So I have no idea where it is or what time or date it is there, exactly. That's odd in the era of the Global Positioning System and clocks that will neither lose nor gain a second during the next two hundred million years; clocks so accurate that we can measure the minute slowing of the Earth's rotation when monsoonal winds hit the Himalaya every year.

It's not so much the name of the day or what the local time is that will affect us when we land in Sydney, however, but the jet lag. Desynchronosis, as it is more formally known, occurs when you override your circadian rhythms by crossing more than two or three time zones in a day. Circadian cycles are our body's accommodation to the rotation of the planet, and no matter if you're sleeping above ground or below it, in constant light or dark, your need for sleep, your body temperature, blood sugar, and appetite, and all your hormone levels will rise and fall throughout a twenty-four to twenty-five hour regimen. Circadian rhythms are present in organisms from the unicellular level to humans, and function whether you actually experience the rotation of the planet or not, so deeply are they buried in our genes. You can, however, induce desynchronosis in both human beings and homing pigeons with lights. With the latter if you offset their rhythm by six hours, they make 90° errors in their flight paths. That piece of trivia is courtesy of science writer Jonathan Weiner in his book on the genetic basis for the subjects in his title: *Time, Love, Memory.*

Weiner is silent on the issue of how such an offset would affect my ability to locate a hotel room, but because our bodies are so wed to this cycle—and our response to our environment is so visual—when nerves in the retina that pass into the hypothalamus provide information about light that's at odds with our genetic instructions, the results are not pretty. The hypothalamus controls the interface between hormones and nerves, hence everything from sleep to appetite. Figure one day per time zone for me to recover from what feels like walking in the land of the living dead. Because I'm crossing more than twelve time zones, I calculate the time difference between Canberra and Los Angeles to know how many days I'll have to wake up at 3:00 in the morning. No matter the difference in days, it's seven hours. A week of trying to keep my face from falling into my lunch plate.

You can't get jet lag from walking from one place to another; our temporal connection to place is so strong that it's possible to sever it only by being in the air. Ironic, when you contemplate the European notion that a God's Eye view comes from above, from an eternity where in theory time is meaningless because it has no end. And that brings me back to the paintings by Aboriginal people and what I'm going to be doing in Australia, which is flying over and driving through its interior while thinking about what it means to paint from an imagined aerial perspective.

Australia is a country where many things are not as they are elsewhere, and a quick primer is in order. Australia is the smallest and lowest and flattest and driest of the seven continents. Only the Antarctic, to which it was once connected, is more remote. The oldest surface rocks in the world are found in Australia, 4.4-billion-year-old zircon crystals from the planet's early crust. Remember that the planet itself is only 4.6 billion years old. Whereas the soils in many areas of Europe and North America were laid down after the retreat of the glaciers 10,000 years ago, soils in Australia can be 2.5 billion years old. Because Australia has had almost no volcanism, and doesn't even receive dust blowing from other continents, it's had no way to replenish its soils, which means all of its plant and animal life has had to adapt to that impoverishment.

Australia used to be part of Gondwana, a supercontinent that included most of the lands now in the southern hemisphere, as well as India and Arabia. Tectonic forces started to pull apart Gondwana around 165 million years ago. Sixty-five million years later Australia started to break away from the Antarctic, and India headed for Asia, where the collision would eventually create the Himalaya. Around 53 million years ago the Antarctic was on its own, which would allow the formation of circumpolar currents and eventually its becoming sheathed in ice. Those two events, the raising of the Himalaya and the creation of the polar ice cap, would help dry out the climate of Australia and force the surviving Gondwana rainforest covering the continent into a slow decline.

Thirty million years before present, a hardy descendant of the Gondwana trees proved itself able to adapt to the more arid climate—an ancestral eucalypt that began moving out of the fringe of the rainforest. Fifteen million years later the continent hit the shelf under New Guinea and raised the mountains there, as well as colliding with the volcanic Sunda Arc of islands, all of which created the rainshadow that diverts precipitation from Australia. With the exception of relatively wet interludes five and then two million years ago, the continent has been mostly dry ever since, and the eucalypt genus evolved into some seven hundred separate species. As the environmental historian Tom Griffiths puts it, no other continent is so dominated by a single genus of tree.

Australia was isolated from New Guinea, its nearest neighbor, by a deep channel, which produced the largest faunal gap in the world; nowhere else in the world did the difference between native species become so marked as between Indonesia and Australia. This is the period, fifteen million years ago, when the first kangaroos were roaming Australia. Initially the size of rabbits, they evolved into animals ten feet tall. The emus were almost as large and the lizards were twenty feet long. This megafauna appears to have died out with the arrival of the first humans, sometime between forty and fifty thousand years ago. It's not clear whether the cause was hunting, or climate change, or both.

Multiple theories exist about when and how the first people came into Australia, but the current state of mitochondrial research indicates that they were part of the migration of *homo sapiens* out of Africa that occurred between seventy thousand and fifty thousand years ago. After crossing Southeast Asia, groups split off to Papua New Guinea and Australia, presumably island hopping for the most part to reach the latter. Forty-five thousand years ago sea levels were 650 feet lower than today; as the glaciers of the last Ice Age melted and released their waters, the sea rose and increasingly cut off the Aboriginal population from significant additions. What attracted those first people is a matter of conjecture; perhaps they were enticed by smoke from lightning-caused fires, proof that there was land across the sixty or so miles of open water. There still exits the possibility that there may have been one large migration or several, of one population of humans or two, although most scientists seem currently to favor the single event. The human fossil record in Australia contains both robust and gracile remains, and it's difficult to tell if they represents two branches of the human race or not. In any case, by thirty thousand years ago most of the continent, except its most arid desert core, was occupied by perhaps as many as one hundred thousand people.

Australia is about the size of the United States, and that doesn't sound like much of a population for such a large place, but 70 percent of its lands are arid or semiarid. Those old and leached-out soils won't support many people, but consider this train of reasoning developed by the anthropologist John Mulvaney, author of the standard text on prehistoric Australia. Assume that from the time the Aborigines arrived during the late Pleistocene, an average of fifty thousand people lived in Australia at any given time, a very conservative figure. That's two thousand generations of people over forty thousand years, or at least one hundred million indigenous people. Mulvaney thinks the actual figure was two or three times greater.

During the last twelve thousand years of Aboriginal occupation, the agricultural revolution was elsewhere spreading around the world, which the soils of Australia wouldn't support, given the lack of a technology to artificially fertilize them. The Australian writer Robyn Davidson, who has been studying nomadism and hunter-gatherers, notes that at the end of the Pleistocene virtually all of us were hunting and gathering, moving from 1 place to another depending on the availability of various resources during each season. By the end of the Middle Ages, perhaps as few as 1 percent of us were still hunter-gatherers by any significant measure. She says that by the mid–twentieth century that figure was diminished to 0.001 percent.

What all this leaves us with are several unique circumstances. Australia is a fragile environment that contains a very deep archeological record undergirding the largest and most intact hunter-gatherer culture on the planet, a culture in a country of less than twenty-one million people of which only 2 percent has aboriginal roots. The majority of Australians arrived within the last 220 years, and they're spread across only eight generations. Fifteen percent of the population is rural, the rest living in five large cities around the coastline. The center of the country, the desert lands once called its "dead heart," is crossed by a single two-lane highway and some substantial dirt roads the Aussies call tracks. It would be as if the United States had a road that reached almost all the way around its perimeter; one paved road from New Orleans to Canada, and Interstate 10 was the only highway crossing from east to west, a two-lane road for most of the way.

Australia appeared to Europeans to be so large and relatively empty and bleak, compared to the temperate environs of Europe and North America, and so newly explored and settled, that airplanes had a great deal to do with the formation of its national identity. Aboriginal dot paintings, those finely patterned abstract grounds upon which there appear to be waterholes and trade routes, as well as mythological lines laid down by the ancestral gods— maybe those environmental factors have something to do with Aboriginal aerial perspective. The distance, both physical and metaphorical, between what we see of Australia from the air and from the ground—and how we see that difference—creates varying perspectives about the nature of landed identity, as the poem by the Aboriginal poet laureate Jack Davis indicates in the lines that open this chapter.

# 3

# DOWN UNDER

# 12. Mount Kosciuszko, 2,228 Meters (7,310 Feet)

*A field all foreground, and equally all background, like a painting of equality. Of infinite detailed extent like God's attention. Where nothing is diminished by perspective.*

—Les Murray

**Everything and nothing is what you see from the summit of Mount** Kosciuszko, the highest point in Australia, and although it is not much more than seven thousand feet high, it's one of the vaunted Seven Summits of the world, the tallest peaks on each continent. I sit out of the wind, my back to a warm rock facing the sun as I eat a peanut-butter-and-jelly sandwich, my staple fare during thirty-six years of mountaineering. Sitting next to me is my sister-in-law, Sue Harris, who since the 1970s has lived with her husband on 350 acres they bought on the west coast of the country outside Geraldton. We're about 120 miles south of Canberra, 2,000 miles east of Sue's home, and happy to be atop a continent. We look out over dozens of other peaks, deep valleys, smoke from farmland fires, and in the distance, ridges blue with atmospheric perspective. It's a view that the early Europeans here found familiar, and even today is so popular with Australians that the government is constructing a toilet into the hillside not far down the trail. It's also a sight utterly untypical of the continent.

Australia consists of three distinct physiographic provinces: these eastern highlands, its vast central plains, and the western shield. The majority of the continent, its interior, is desert: arid mountains, red sand dunes, stony flats that look more like another planet, and white playas. As I said earlier, Australia is the flattest continent with an average elevation of only 1086.6 feet (333 meters). But, as all continents must, it has a highest point, this rounded peak christened Mount Kosciuszko in 1840 for a Polish general by the explorer and geologist Count Paul Strzelecki.

Our perch today is a bump within the Great Dividing Range, a complex of peaks, ridges, highlands, uplands—just generally raised terrain—that runs for almost 2,300 miles in a great arc extending from Cape York on the northern coast clear to Melbourne on the southern. It parallels the entire eastern seaboard of the country, and defines the watershed of almost half the continent. Rains falling on the east run into the Pacific Ocean, those on the west, drain into the Murray-Darling Basin. The basin, which is itself so flat that the Murray and Darling rivers tend not to flow so much as pool from one depression to another, receives only around 6 percent of Australia's scant rainfall, but produces 40 percent of the country's agricultural products. This necessitates 70 percent of the country's irrigation to be deployed in the basin, which is in increasing conflict with urban water usage, mining activities, and the needs of the environment. From my high point here and during the next two months I'll work my way down into that and other interior basins, walking and driving and flying farther and lower inland, until I'm below sea level on the playa of Lake Eyre, the infamous "dead heart" of the country.

But today I sit with Sue on Mount Kosciuszko, a peak in what's called rather grandly the Australian Alps, and at the head of the Snowy Mountains. Uniquely in all the world, because Australia is the only continent that is also a country, this summit is also a pinnacle of national geographical identity. And that makes it an ideal place for me to begin an aerial survey of Australia, a country where European representation of its landscape progressed in a relatively short time from sea level to the aerial. Because Australia's settlers arrived in 1788, the opening of the country to European eyes coincided, as did that of the American West, with the twin developments of aerial and photographic technologies. In turn, those tools had a profound effect on how the country was represented graphically, from fine art to advertising, hence in politics. Given the country's aridity, thus its vulnerability to changes in climate, politics in Australia is inevitably about the environment. The synergy among the aerial, the arts, and the governing of the country is as, or more, apparent here than anywhere else in the world.

The settlement of Australia also occurred as Alexander von Humboldt was exploring South and Central America, and as his mountaineering exploits were becoming fashionable to emulate. It was a passion that spread first among explorers and geographers, but increasingly among artists as well. Hence Count Strzelecki's ascent of Mount Kosciuszko. According to signage along the metaled track leading here, the shape of the summit reminded Strzelecki of the national Polish hero's tomb. I am having trouble with this piece of folklore, given that the name was first applied to what was originally thought the higher peak, now known as Mount Townsend, which is just to our north. What is now named Mount Kosciuszko is just a few inches under 7,310 feet; Townsend, despite its more rugged appearance, tops out at only 7,247 feet. The New South Wales Lands Department simply transposed the names when the error was discovered shortly after the dawn of the twentieth century. This causes me to sigh as I attempt to figure out where the artist Eugene von Guérard was sitting when in 1862 he walked up here to sketch a panorama of the Snowy Mountains. Maybe it's my lingering jet lag.

Yesterday I'd gone over to the National Galley of Australia in Canberra to look at von Guérard's wonderful little oil painting, *North-east view from the northern top of Mount Kosciusko* (Plate 7) done the year after his ascent. (In addition to rectifying the elevations, the Australian government in 1997 added a z to the name to align the spelling more closely to the Polish). In the left foreground and across the bottom of the image a mass of dark gray granite boulders frames the view, which falls away into the distance. Rocky crags march up a sere ridge in the middle ground, and in the background other peaks range across the horizon. Dark clouds presaging rain drift in from the upper left while a man standing on a flat rock gestures grandly at the scene. Several smaller figures conduct their business about the summit, Guérard having constructed a theater of science. The man waving his arms at the mountains was none other than the noted meteorologist and geologist Georg Balthazar von Nuemayer, who had been sent off by no other than Humboldt himself to establish a geomagnetic laboratory in Melbourne. Nuemayer started at sea level and worked his way up to what he thought was the highest point on the continent in order to make his observations, and in the tradition of colonial explorers everywhere, took with him the finest landscape artist the land had to offer.

So here I am, sitting on the summit with my sister-in-law, whom I last saw thirty-six years ago shortly after she and her husband and two children emigrated to Sydney. We've finally managed a reunion, which we're celebrating by walking up a summit she's always wanted to obtain, the highest point of her adopted homeland. And "Kozzie," as it is more familiarly known to Aussies, who with great friendliness abbreviate even the name of their own country to "Oz," is a wonderful place from which to assay the history of mountain climbing, which is its own kind of aerial addiction. The aptness of this particular peak is due in part to the fact that Kozzie is the smallest of the Seven Summits sought after by peak baggers the world over. Kozzie suffers, however, from the indignity of being mocked as a member of the club because its modest elevation is far shorter than the next one in elevation, the Vinson Massif in Antarctica (16,050 feet), not to mention Mount Everest, where the actual climbing toward its 29,029-foot summit doesn't even begin until above 18,000 feet.

The idea of the Seven Summits sounds like a marketing scheme and it was no surprise to the climbing community when it turned out to be the wealthy American businessman, petrogeologist, and amateur mountaineer Richard Bass who declared, upon his reaching the top of Everest in 1985, that he was the first person to ascend the highest points on all the continents. His book with its eponymous title appeared the next year and quickly prompted the man who was unarguably then the world's greatest mountaineer, Reinhold Messner, to name New Guinea's Carstenz Pyramid (16,024 feet, and unlike Kozzie a genuinely difficult summit) to be the highest on the Australasian continental plate. It gets progressively sillier from then on—such as, who could climb the original seven in the shortest time period—Mastan Babu of India, 172 days—and every variation of the seven or eight summits imaginable. (If you're that curious, you can go to 7summits.com for the rolls.)

Although I've climbed mountains on several continents and led treks to the base of Everest, Kozzie is the only one of the Seven Summits I'm likely to attain, and I'm quite pleased with this jumble of granite boulders among which the two dozen or so of our fellow hikers drink water, eat sandwiches, clown for cameras, and in general celebrate in individual ways the joy of getting atop anything that has the superlative "highest" attached to it. Conventional wisdom has it this was not always the case among humans.

When I was a young climber reading about the history of the sport, it was a given that human beings for the most part avoided mountain summits until the Italian poet Petrarch walked up Mont Venoux, a 6,263-foot mountain near Avignon in 1336. But even he was quoting Livy to excuse his curiosity, his predecessor writing that the Macedonian King Philip climbed Mount Haemus as part of a military campaign, which would have been around 336 BC. Climbing in Europe was not established as a desirable modern activity, however, until Horace-Bénédict de Saussure, a Swiss natural historian, visited Chamonix in 1760 and offered a prize to the first person to ascend Mont Blanc, at 15,774 feet Western Europe's highest peak. Even after that event, the superstitious local farmers continued to avoid the mountain, seeing it as the home of angry gods tossing about thunderbolts.

Mont Blanc was finally climbed in 1786 by a local doctor, Michel Paccard, and a crystal hunter, Jacques Balmat. Since then the history of climbing has been construed by the public, and many climbers, as a never-ending race to be first up anything by the hardest route possible. When climbers ran out of peaks, they started on individual cliff faces. Then buildings. But I found it difficult to believe that such a universal desire to climb was a modern preoccupation, and several things conspired to alter my ideas about the nature of climbing mountains. First, everywhere I went there was evidence that people were functioning at high altitudes among the peaks, whether it was neolithic rock arrangements on 10,000-foot summits in the Great Basin, or prayer stones on Himalayan passes at 18,000 feet. And then, in 1991 the frozen mummy of "Ötzi the Iceman" was discovered melting out of a glacier on the border of Italy and Austria. The body, dated to about 3300 BC, was found at 10,500 feet and clothed in a bearskin cap and waterproof shoes meant for traversing snowfields. Since then numerous other bodies have been found frozen in glaciers around the world. It became obvious that people were traversing passes, exploring mountains, and fighting in the mountains over territorial claims for thousands of years. They may not have been scaling the toughest peaks, but they were seeking the high ground for any variety of reasons, sometimes to send signals by fire over long distances, sometimes for ritualistic and symbolic purposes—and I cannot help but believe sometimes just out of curiosity and the joy of standing on top and looking around. Neither Petrarch nor Philip invented the impulse.

Back to Mount Kosciuszko. Among our fellows summiteers today are two German girls in shorts and cotton tops, a family of rural Australians with grandmother dressed in street clothes and tennis shoes, two elderly people in matching white hats, and a student in dreadlocks and a serious mountaineering parka. The first Europeans entering the Australian mountains in the 1820s were guided by local Aborigines, who had been making paths up here every summer since the end of the last Ice Age in order to feast on the protein-rich Bogong moths that flew to the mountains to escape the heat below. No one we see today looks to have any Aboriginal heritage.

An estimated thirty thousand people visit the summit each year, the most popular route being from the top of the ski lifts operating out of the resort town of Thredbo, and the way we'd chosen to take. The town is a three-hour drive south of Canberra, and sits in the lower reaches of the Kosciuszko National Park. The first land set aside in the Snowy Mountains for public use was 62 square miles for the preservation of game in 1906; three years later a road was completed to the top of the peak. The preserve was enlarged into the 1,950 square miles of Kosciusko State Park in 1944, re-designated a national park in 1967, and today consists of about 2,683 square miles. The road wasn't closed to vehicle traffic until 1974, when the lack of parking spaces became a problem.

Once the dusty road from Charlotte Pass was closed, Thredbo became the trailhead of choice. The first ski resort there opened in 1957, and the ride we took this morning on the double chairlift lifted us up 1,550 feet in just over a mile. Although the ski runs are due to open in a month, there was no evidence of snow anywhere. We rode past dozens of the 135 hydro-guns that are supposed to be making snow starting this next week, but it's way too warm. Nothing would stick. Thredbo has the largest snowmaking operation in the southern hemisphere, able to run most of its snow guns simultaneously. The 2,905 gallons of water pumped per minute should translate into a total of 15,891,600 cubic feet of snow. That's especially critical on the lower runs, which in some years hold almost solely artificial snow.

Snow in these mountains feeds two major river systems, the Snowy River and the Darling, and given that Australia is in the midst of what some scientists have defined as a thousand-year drought, the use of the water here to make recreational snow is controversial. I know the argument goes that the artificial snow melts and heads back downhill, but given the energy used to pump it up here, and the gas burned by skiers driving uphill to get to Thredbo, it's a contradictory idea to be making snow during drought years abetted by global warming. An environmental report on the park published in 2001 estimated that even in a best-case scenario the snow cover in the Australian Alps will decline 18 percent by 2030 and 39 percent by 2070. The worst-case scenario is a 96 percent reduction by that latter date. Global warming is one of those enormous disasters to which we instinctively want an equally large, singular solution, but that's not the way it works. We create the disaster one car at a time, one coal-fired electricity plant, one cement plant, one airplane ride at a time, and it will have to be solved the same way. Sue and I make no attempt to justify the fact that we're participating in the situation, although we had allowed ourselves clucking tongues at the large American SUVs in the parking lot next to our compact Toyota. Hypocrisy over these issues is as gently inclined a slope as the walk up Kosciuszko itself.

When we jumped off the chair at the top we were just below treeline, the snow gums ending around 6,500 feet, and the temperature at 11:00 in the morning in the high 40s Fahrenheit: comfortable walking for long-sleeved t-shirts, and we shouldered our daypacks. After we climbed a steep few hundred feet past the top of the ski lifts, the trail leveled off onto paving stones for a few hundred yards, then became a walkway of continuous metal grating. Construction of the slightly elevated walk took seventeen years and it was finished in 1999. It was surprisingly comfort-able to walk on, in that it gave slightly, but more importantly, it kept us from treading on the fragile alpine meadows for the next four miles. At one point I stopped to take a photo of the old trail, which was running parallel to the newer path. After twenty-five years the original track was still a three-foot-wide rut under recuperation.

Every few hundred yards interpretive plaques stood next to our walk-way, placed there to draw attention to rare and/or endangered plant and animal species. Sue was thrilled to learn that a tiny marsupial lives on the mountain. Known only through fossils and thought to be long extinct, the mountain pygmy-possum was rediscovered by accident in 1966 when some skiers at a hut on a nearby peak adopted one living inside, then took it back to Melbourne for identification. Only 2,300 or so of the mouse-sized marsupials live in colonies on the highest peaks here, and because they are adapted to hibernate during winter (the only marsupial to do so), they haven't yet been bred in a zoo. I, on the other hand, am enamored of the equally petite varieties of climbing galaxids found in the streams. The almost invisible fish are being harassed by trout, which were introduced as a game fish during the last century, but are not as endangered as the marsupials. The fish, which grow larger elsewhere but at this altitude remain tiny, are one of the few species able to come out of the water and express mobility, most often by using their fins to squirm up rock faces and even dams to pools of water. Global warming will wipe out the pygmy-possums by destroying their habitat, and the galaxids won't use this as a home anymore.

After walking up a small rise and past the lookout from which Kosciuszko was visible for the first time, we traversed the headwaters of the Snowy River, which at that point were nothing more than trickles of water slowly being released by patches of sphagnum moss. Those drips and seeps aggregated into rills the width of my wrist and disappeared over an edge to the south. Further down the small bowl they reappeared as actual streams before dropping off into the valley below. Underneath the grating the alpine grasses were still green, although small edges of ice bordered the shadows. It had been cold enough at night to freeze, but not staying cold long enough for the ground to really chill and hold snow.

A second lookout was built on a bump toward the far side of the bowl, and an overlook faced east over what one of the plaques says is the only glacially carved lake on the continent, Lake Coolapatamba. More a silted up pond, the tarn is one of four alpine lakes in the region around the peak, and I was happy to tell Sue that it was also my older son's birthday. His name is, of course, Tarn, named when I was still learning to climb near the small lakes at the eastern entrance to Yosemite National Park. The last permanent ice in Australia, a small ice cap about twenty-three square miles in extent, melted here twelve thousand years ago.

The metal way ended where the old road from Charlotte Pass joined in, and a construction project was underway. Pickup trucks were parked near a concrete bunker with glass windows being built into the side of the hillside, the dirt roof pierced by large ventilators. That would be the toilet. So many people now visit the summit that it was beginning to smell like a urinal, so a pump-toilet station was being installed. Hay bales surrounded the construction site to prevent erosion, and much of the land around the junction was carpeted in a coarsely-woven organic blanket with new plants poking out of it. All of this to mitigate the human desire for an aerial view.

After that the hike adopted the remnants of the old dirt road, which wound around from the western side of the mounded peak to the south and east before reaching the pylon marking the summit. And here we sit looking north to the cliffs of Mount Townsend, its more rugged profile a clear inducement to the nineteenth-century explorers to name it the taller peak. It simply offers more hints of the sublime than does the less imposing Kosciuszko, which was enough to distract the gaze of both the scientist and the artist.

It's ironic that the view from here has so little to tell us about the nature of the continent as a whole, but that is more often the case than not from mountaintops. The view from the Dolomites in northern Italy won't tell you much about Tuscany, for example. But humans are always seeking an expanded view, a picture of the world that is deeper, broader, more inclusive, more incisive. The painting by Guérard displays mankind as standing above daily life, engaged in what was construed by predecessors such as the painter Caspar David Friedrich to be a deeply romantic and heroic endeavor: aloofness. That is to say, by remaining distant from everyday affairs, one might have the mental and emotional capacity to grasp the sublime nature of the universe. I suspect that such a notion wasn't invented by eighteenth-century Europeans, but is rooted in hardwired cognitive behaviors arising from survival skills, as is the longing to make panoramic images, as well as aerial images. It's all about the larger view that gives one more information, that allows one to see farther and thus predict what lies ahead.

The history of aereality in Australia is both similar to and radically different from that of North America. The European attitudes and technologies deployed in exploring and settling the last two great land masses, Australia and the American West, were similar enough to produce almost parallel art histories, and in 1998 the National Gallery of Australia in Canberra and the Wadsworth Atheneum in Connecticut organized a landmark exhibition celebrating the fact. "New Worlds from Old" compared and contrasted nineteenth-century landscape painting in the two countries, and the predilection for the wider, larger, higher view is present in both countries.

It's not my intent to attempt to summarize all of Australian art history, be it Aboriginal, Colonial, or contemporary, but I'll touch on all of them in order to understand how persistent and varied the aerial view of the world is embedded in humans. And it's easiest to start with the European topographical tradition of the elevated panorama. It has a common root between the two countries with Alexander von Humboldt. The nineteenth-century scientists and artists exploring the two worlds were inculcated with his ideals, ideals that necessitated climbing to heights in order to form a connected, unified theory of the environment. Just as the citizens of the United States thought a river must flow from the Rockies to the Pacific Ocean, so the antipodean settlers hoped to find a great central river in the interior, something along the lines of the Mississippi, or perhaps deep bodies of freshwater comparable to the Great Lakes. Both groups would look in vain, dissuaded only when explorers took to the high ground. In the case of North America, it was John C. Frémont climbing up the Wasatch and the Sierra, looking inward from each and finding the fully enclosed Great Basin. In Australia, it would eventually be Lake Eyre the Australians found, another landlocked basin and one of the world's largest intermittently dry lakes.

The first European pictures made of Australia, like those of the Antarctic with which it was usually conflated, were imaginary ones. Pythagoras in the sixth century BC and other Greeks, such as the geographer Crates in the second century BC, proposed that southern landmasses were necessary to balance the known world of the northern hemisphere, lest the planet roll over. Which, when you think about it, was essentially an aerial view of the planet. Although there is no direct evidence supporting the claim that Portuguese sailors actually set sight on Australia during the early 1500s, the myth of such a continent continued to propagate throughout European cartographic circles. But the first confirmed sighting of the continent was made by sailors aboard the Dutch vessel *Duyfken* in 1606. Unfortunately, no artist was aboard to record a view.

Spain's King Phillip II had initiated the tradition of artists accompanying scientific expeditions when he sent Francisco Hernandez to Mexico from 1571 to 1576. Among his other duties, Hernandez was charged with collecting specimens of plants, and he hired three Mexican artists to portray both the plants and animals of this New World. It was a tradition that the British would adopt when sending James Cook on his circumnavigations of the Pacific a hundred years later. The *Duyfken*, however, was not on an exploring voyage, but looking for gold. An artist was the last person who would have been aboard, and the first drawing of the continent based on observation appears only as a fragmentary outline of a small section of the coast west of Cape York on a chart of the voyage.

The first actual picture of the continent was a coastal profile done along the west coast by Johan Nessel in 1658. Coastal profiles, running panoramas drawn of coastlines, had been developed by Portuguese navigators sailing down the relatively featureless west coast of Africa during the 1400s as they sought a way to establish commerce with the Near East and Asia without sailing through the Mediterranean. The profiles, when lined up with shoreline features and linked to soundings, allowed subsequent navigators to avoid shoals, and locate safe anchorages and freshwater. The technology moved north as Jews fled the Inquisition, and was adopted by the Dutch in the late 1500s, hence Nessel's almost four-foot-long profile.

The artist and cartographer Victor Victorszoon, sailing with the de Vlamingh expedition of 1696, made eighteen watercolor views of the continent's west coast, and in addition to being the earliest consecutive profiles of its coastline, I consider them the first pictures of Australia to display an elevated viewpoint. Fifteen of the paintings are exquisite coastal profiles that often display the continent's outline in a rare manner, curved and slightly elevated as if seen through a fisheye lens mounted high upon a mast. The effect is to make the viewer feel almost as if in motion along the coast, a deck rolling underfoot, with glimpses of the interior behind the coastal hills about to be revealed as the ship crests a wave.

From 1606 until James Cook arrived on the east coast of the continent in 1770, more than fifty European explorations sailed around or landed in Australia, many of them making charts and coastal profiles. It wasn't until Cook brought Sydney Parkinson ashore during the captain's first circumnavigation about the Pacific that an artist had a chance to sketch the landscape. Once the British had realized how useful the technology was, they hired Dutch artists to teach the Navy how to draw and paint. Cook, a gifted surveyor and chartmaker, was among the personnel trained in making profiles, and he often worked with his artists in assembling views of newly discovered lands. Cook was the first European to sail up the east coast of Australia, and Parkinson made coastal profiles and sketches as they did so. In May of 1770 Cook sailed into Botany Bay and made a map of it, one of those aerial landscapes that outlines the coast and represents the interior with a scattering of small trees to indicate forest. The first elevated views made in Australia from the ground were done of harbors, looking back at the ocean and thus connecting them to the coastal profiles and charts.

The last great maritime voyage of exploration to Australia was that of Matthew Flinders in 1801, when his ship became the first to circumnavigate the continent. With him was the topographical painter William Westall, who made meticulous coastal profiles. As Bernard Smith points out in his landmark study, *European Vision and the South Pacific*, Westall's profiles not only enabled subsequent sailors to pilot their way safely to anchorages even during storms, but also captured specific geological strata and plant species. The twenty-one-year-old artist was also a romantic in search of more exotic fare. Not content to make topographically accurate pictures from a distance, he was constantly clambering up hills along the shoreline to make ink-and-wash sketches. In 1802 he sketched Port Bowen from the top of a gully where the crew collected freshwater, his drawing the basis for a large oil he painted in 1811. It transforms a foreign land-scape into a soft atmospheric study, carefully delineated trees in the fore-ground framing the scene as it falls away to the bay far below. In the foreground to the left of the trees, and just below the vantage point of the artist, are three Aborigines returning from a successful hunt, figures that add scale and ethnographic focus.

Likewise, the first representations of Sydney Harbor were painted within a few years of the first European landing. The convict artist Thomas Watling arrived in Sydney in 1792 and within two years was pro-ducing views of the harbor from the surrounding heights. One published in 1804 was typical. Composed as seen from the hills along the cove's western shore, the elevated vantage point allowed him to display the rela-tionship of the settlement to the bay and the growth of the town, and to counterpoise a group of Aboriginal people onshore against the technolog-ical superiority of the new ship floating in the calm waters just offshore. The European image maker may have been a reject from Great Britain, but his point of view is higher than that of the indigenous people, his gaze encompassing them and establishing an immediate hierarchy, as with the Westall painting.

Sixty miles west of Sydney and walling it off from the interior of the continent were the Blue Mountains, part of the Great Dividing Range. Not really mountains, but a steeply eroded sandstone plateau surrounded by cliffs, the first road was not constructed through the range until 1815. This most prominent vertical feature attracted artists, who used the cliffs to establish an early Australian tradition of the sublime. John William Lewin went along with Governor Macquarie in 1815 to celebrate the road in a series of watercolors, and the party marveled at the views it afforded of the countryside below.

The French artist E. B. de la Touanne was making sketches in the moun-tains in 1825, but it was Augustus Earle, the American-born and legendary wanderer who had arrived in Sydney that same year, who would create the first notable painting of its scenery. Early in 1826 he crossed the range to Bathurst, sketching cliffs and streams in preparation for *Waterfall in Australia*, the most reproduced early view from atop the plateau. In a variation on a theme, he portrays himself in the foreground as seated below an Aboriginal guide, who he has posed standing upon a ledge as a somewhat heroic figure, very much in the mode of the Noble Savage. In 1827 he created eight views of Sydney from Palmer's Hill, the basis for a panorama exhibited in London the next year, an elevated, enlarged view that made comprehensible to the home country how the colony was progressing.

The road through the Blue Mountains marked the initiation of Australians into the interior of the continent, and while early bird's-eye views of cities such as Melbourne were being made, surveyors were beginning to profile the Murray-Darling Basin. Sir Thomas Mitchell, Surveyor-General of New South Wales from 1826 until 1855, was but one example, albeit perhaps the most accomplished, of a generation trained not just as mappers but also as topographical draftsmen and landscape painters, much in the spirit in which the British sailors had earlier been taught. By necessity, surveying requires that elevated vantage points be assumed in order to sight from point to point, especially in rugged terrain. Mitchell crossed the Blue Mountains, mapped the Darling and Murray rivers, and discovered the Grampians, which are the tail end of the Great Dividing Range.

Mitchell made panoramic drawings of the terrain he was busy converting into territory during his four expeditions conducted from 1831 to 1846, great elevated sweeps of vision such as *View from Jellore Looking North and North-West over the Unexplored Interior of New South Wales*. A lone Aboriginal figure sits on the edge of a precipice with his back to us, gazing out over seemingly empty hills and valleys stretching to the horizon. That is, of course, a failing of the elevated view. The country had been thoroughly settled by indigenous people for millennia, a fact apparent only upon closer inspection at ground level. Twenty years after Mitchell had surveyed New South Wales, von Guérard was publishing his popular views of the countryside, including the one from Mount Kosciuszko, a book that included topographical and travel information for the enlightened tourist.

Any view, no matter how expansive, panoramic, elevated, or aerial, carries the frame of its maker, which in turn is determined by the culture in which the artist is nurtured, but also by the longer-term predilections established in our biology by evolution. The human species didn't first evolve in Australia, but on the savanna and in the broken woodlands of Africa. Our vision—which includes not just our eyes, but our body-wide neurophysiology and brain—is biased toward that environment, and we have trouble seeing others, as I wrote while flying over the American Midwest. It's harder for us to know how to measure ourselves, to establish distance and scale, in a landscape without strong vertical elements, without tree limbs against which we sight our own limbs. That is, we're tree climbers even when out roaming the desert. And most of us tend to avoid deserts as permanent habitat when possible not just because we're too hot in them, or thirsty, but because we're visually discomfited.

The early Australian settlers looked at what they thought was almost an English parkland in places (actually a habitat created, at least in part, by Aborigines using a fire regime to increase productivity designed for human consumption). They thinned the forests as necessary to make it even more so, and when travelers went over the Blue Mountain road to Bathhurst—into the Murray-Darling watershed—they found a land that today still looks at first glance like the rolling hills of England. Eugene von Guérard was trained to paint landscapes that expressed those ideals and their sublime counterparts, the waterfalls and mountains of Europe. That's what he sought out in Australia, it's what his patrons wanted, and it was only explorers looking for what they thought would be a wet and lush interior who kept pushing into the deep arid center of the continent.

It would be the artists accompanying those exploring expeditions that would be the first ones to picture the Australian desert, people such as Guérard's friend Ludwig Becker accompanying Robert O'Hara Burke's expedition seeking a route to the northern coast of the country. Whereas Guérard's paintings and lithographs helped to fix firmly in the Australian art tradition the elevated view, a rural complement to the urban bird's-eye views, the desert art would emphasize from ground level what seemed to be to European eyes a never ending and flat wasteland.

By the time I finish contemplating the role of Guérard in the formation of aereality in Australia, it's time for us to begin the hike back to the top of the chairlift, which will close in mid-afternoon. I've always thought it a luxury the few times I've taken a chairlift down a mountain, and today is no exception. Sue and I swing our legs and enjoy the sensation of flying over the valley, sometimes close to the slope, sometimes hundreds of feet above it, over gullies filled with snow gums. There's a new sport being developed at ski resorts around the world, zip lines that you clip yourself to with a seat harness and fly down to far points at up to 60 miles per hour. It seems as if humans will invent anything to obtain the sensation of being a bird, to soar above the ground as if unfettered by our feet of gravity.

# 13. Pennyroyal

*"When you fly," the young man said, "you get a feeling of possession that you couldn't have if you owned all of Africa. You feel that everything you see belongs to you—all the pieces are put together, and the whole is yours; not that you want it, but because, when you're alone in a plane, there's no one to share it. It's there and it's yours. It makes you feel bigger than you are—closer to being something you've sensed you might be capable of, but never had the courage to seriously imagine."*

—**Beryl Markham**

**The entrance to the gorge is narrow and approaching fast, a situation** pilot George King addresses by simply laying the Cessna over onto its left wing. The painter Mandy Martin in the backseat, myself in the copilot spot, and George all gaze calmly out the port windows as the shadowed bottom of the gorge a few dozen feet below rushes by the wing tip. I feel like I'm about to fall out of my harness. Then the rocky surface drops away beneath us by several more yards. "See the waterfall!" George exclaims, a modest trickle of water falling down to the creek. "Cool," I reply—but, isn't George supposed to be looking out the windscreen? He's still peering down when I glance up to discover the narrow walls ahead of us bending sharply to the left. Before I can say anything he's adjusted the controls and we slip around the curve to exit the basalt slot. He rights the plane just in time to clear the eucalypts.

We're still less than a hundred feet off the deck and flying at tree level over a pasture on George's property. He points out how healthy the canopies of the gum trees are—and of which we have a startlingly close view—and how well-watered his cattle are, the cattle underneath us who don't bother to look up, apparently used to their owner's flying habits. I can see individual furrows made during the 1940s that he's left fallow. "See the grass growing in the bottom of the gullies? The land's starting to heal." I'm counting blades of grass. He says this as we circle right, jump over a set of low telephone lines and fly right into a hill, George pulling up the nose at the last second. The land flattens magically before us and we set down on his dirt airstrip. As we pull into the hangar halfway down on the left I discern a severely bent prop next to the plane. George shrugs. "I was mustering cattle. You know, you fly pretty low." There is no trace of irony whatsoever on his tanned and smiling face.

George King, a young Australian grazier, has been living and raising cattle here under Mount Macquarie since 1996. Along with Mandy Martin's husband, Guy Fitzhardinge, he's one of the more outspoken advocates in the Murray-Darling Basin for sustainable agricultural practices. Guy, a trim but stocky man in his late fifties, owns 7,500 acres on three parcels in the vicinity, and grew up in a small town nearby. He and Mandy live on the property known as Pennyroyal, a handsome spread of rolling hills, creeks, and gum trees that he's turned into an officially sanctioned nature preserve even while it remains a working ranch. His viewpoint, as a grazier from the older generation, is that "In order for this land to be healthy now, because people have been in it for so long, people need to maintain it." George is of the next generation, and represents what Guy knows to be the only hope if agriculture is to remain viable here. He also knows the presence of the younger man and others like him is far from a guarantee. It's not just the drought and the struggle over who needs the water more, the farmers or urban dwellers downstream. Since the basin was settled in the nineteenth century, people have ripped out more than fifteen billion gum trees. Not only has that caused severe erosion problems—easily visible from the plane—but the shallower roots of the crops replacing those of the trees has allowed immense amounts of salt to mobilize in the groundwater. Driving in the basin can resemble a rollercoaster ride as they can't repair the roads as fast as the salt heaves them up.

Mandy and I take our leave of George, and lower ourselves gingerly into the seats of her all-wheel-drive Audi station wagon. It's not until we get back on the road that we look at each other and start laughing. Mandy's even less fond of small airplanes than I, and it's been a lesson in how much we both trust George to have not been whimpering in our harnesses as we flew half upside down through the gorge.

I've driven up two and a half hours from Canberra, the capital city of Australia, to stay with Guy and Mandy for a few days, the first of what I hope will be many visits. They're a unique couple. Guy is not only a grazier, but is close to finishing his Ph.D. in environmental history. Mandy is one of the leading contemporary painters in the country, and has led numerous field trips with other artists, writers, and scientists "out bush" to produce one of the most significant bodies of environmental art in the country. Last evening, as we turned off the two-lane highway that leads up into the hills east of Cowra, and started into the eucalypt forest, we were surrounded by kangaroos bounding alongside the car. "The welcoming committee," Mandy explained. "We have to either speed up or slow down as there's a break in the fence coming up, and those ones on the left will cross the road to go through and onto our property." Sure enough, we slowed, and they loped across.

The house, tucked down on the north side of a hill, looked small at first, but it slopes ingeniously down the hill under a slanting roof. In addition to a large kitchen and various dining areas, it holds four bedrooms, an office for Guy and a large studio for Mandy. Oriental rugs soften the brick floor, and various nooks are kept private from one another with walls covered with art and bookcases. If I had to pick a single palette to describe the place, I'd say rustic Italian, despite the clerestory windows and modern lines of the roof. Guy built the first part of the house in 1976, and then after meeting Mandy added on the master bedroom, a second guest room, and Mandy's studio.

When I rose this morning just before dawn, I rolled over in bed and stared straight out French doors and under a wisteria-laden trellis into softly lit hills. It's mostly all grazing land that's been cleared, each lineament of ridge and swale defined by gum trees. An inch and a half of rain a couple of days ago had greened it up temporarily, and it was as pleasant a pastoral scene as I could have imagined. Pennyroyal is four hours west-southwest from Sydney and on the eastern edge of the Murray-Darling Basin, which drains almost all of New South Wales west of the Blue Mountains. The property is in the central tablelands, a district that's always been among the richest and most productive land in what is Australia's breadbasket. Cows graze peacefully under gum trees, the branches of which hold brilliant green Superb Parrots. It's as strange a combination as you can imagine and not a little deceptive.

This is land that people have been altering for forty thousand years, first with hunting and gathering abetted by deliberate firing of the flora, then with deforestation and irrigation schemes during the nineteenth century, and more recently with industrial machinery and phosphates. In most cases the basin has been used in unsustainable ways for 150 years, and now is threatened with urban developments. The drought is so severe that the suicide rate among farmers has risen dramatically as their livestock either die of thirst or are sold off, and many owners are selling out to corporations with plans for second-home subdivisions.

The Murray-Darling Basin provides upwards of 40 percent of all Australia's agriculture, and is the largest river basin in the country. The Murray, which runs west from the Southern Alps and through the bottom third of the basin, is the seventh biggest river system on the planet. The Darling, which lies on the west side of the basin, runs more than 2,100 miles south from Queensland and drains most of New South Wales. The basin covers 15 percent of the country and is the size of France and Spain combined, a shallow dish just a few feet above sea level that was once an inland sea. For the last four million years slow rivers puddling along have very slowly carved the basin into a complex web of channels that sometimes flow as little as 1 percent of their mean annual delivery, but sometimes at a hundred times the mean. As I mentioned while flying from Los Angeles, Australia has the most variable climate on earth.

The recent drought here has been going on for seven years and is so bad that the government is on the verge of declaring all the basin's water reserved for drinking purposes. If they don't change the agricultural pattern in the basin soon, it's going to cost the country about $16 billion a year in lost crops alone. Only 6.19 percent of the country is arable, and this would be a huge blow to the Australian economy, which is already importing more expensive fruits and vegetables to compensate for the drought. If conditions continue to worsen, which global warming threatens, then Guy and Mandy would have to sell their cattle and become city dwellers, and the Superb Parrots, already classified as a vulnerable species due to decreasing habitat, would be further threatened.

On the far horizon, fifteen miles distant and just visible from my bedroom as a low plume of dust, was the Cadia Hill gold and copper mine, the second largest open-pit mine in the country. Mines use enormous amounts of water to process ore, and Cadia, which started operations during the wetter times of the 1990s, is still expanding during the drought. They've started pumping water from the Belubula River adjacent to Guy and Mandy's property, and are now negotiating with the nearest town, Orange, to buy some of its drinking water. Given that the mine is the largest employer in Orange, the outcome is easy to guess. Because Australia is so ancient geologically, it has a wealth of minerals, but much of it is buried under hundreds of feet of regolith (the unconsolidated rocks lying atop bedrock). It sends overseas so much of its unrefined bauxite, uranium, coal, gold, copper, and iron that the country is often referred to in the international marketplace as a quarry. Eighty percent of its export earnings come from minerals that, increasingly, are being sent to China. Those dynamics are difficult to ignore if you're sitting on a city council.

Visible from my window this morning, then, were some of the major factors competing for Australia's scarce water. Drinking water versus agriculture versus mining. Guy tells me this is "grass and white box" country, an ecological community listed as threatened, which used to stretch all along the western slopes of the mountains from north in Queensland to South Australia. Only 1 percent of it is left now, and Guy has been replanting native trees here for years, as well as fencing off stream banks from his livestock to control erosion and hang onto water. He long ago sold off his sheep to concentrate on raising beef, a much more sensible animal for the country. But despite these and a host of other measures to increase sustainability, it's obvious that he along with all his neighbors is dipping into water and feed reserves to keep cattle on the land.

Rising earlier than Guy and Mandy—still off kilter from local time zones—I wandered out into the living room where an enormous chunk of said white box was smoldering in the fireplace. Once autumn arrives, Guy lights a fire that burns slowly all winter long. I settled into a corner of a couch nearest the warmth, turned on a reading light, and decided I'd be happy to live here for as long as the water holds out.

The day eventually beckoned, however, and after breakfast Guy was off to cull his pregnant cows. They don't do well in a drought, and this might have been just the first in a series of sales he would have to make, depending on precipitation this coming year. Mandy and I took the opportunity to talk in her studio and I began my firsthand inspection of her work, which I had been anticipating ever since being introduced to it two years earlier by our mutual friend, the American environmental scholar Scott Slovic.

Mandy's paintings, in a nutshell, are large landscapes dealing with exploration, be that defined as heroic exploits out bush during the nineteenth century, or the fieldwork of contemporary anthropologists, or the intersection of Aboriginal Dreamings with contemporary terrain. She often paints with pigments ground from rock collected in the site depicted while she's sketching, in particular the incredibly varied ochres that run from gray and green through yellow and more shades of red than we have names for. She sometimes includes words from earlier artists or explorers, and her artistic heritage includes masters of the European sublime Salvator Rosa and J. M. W. Turner, as well as Australian painters William Piguenit and, perhaps most importantly, Ludwig Becker.

The seventeenth-century Italian painter Rosa painted turbulent landscapes that prefigured the birth of Romanticism. Turner, painting during the next generation, melded an acutely accurate topographical sensibility based on field observations with the ability to deconstruct a scene into almost totally abstract plays of light. Both artists, by providing ways to construe nature on the large scale, have been important in the history of exploration, and as a result our ideas about the environment. Meriwether Lewis invoked Rosa's name during the Lewis & Clark expedition across the American West, and Turner was an inspiration to successors such as Thomas Moran in Yellowstone, Edward Wilson in the Antarctic, and Thomas Cole in the Hudson River Valley. Piguenit, a Tasmanian artist and friend of Guérard's, accompanied numerous expeditions through Tasmania and New South Wales, and was perhaps the foremost Australian painter of the sublime during the mid- to late nineteenth century.

Ludwig Becker may have been the finest artist in Australia to accompany an expedition during the nineteenth century, as the art and environment scholar Tim Bonyhady has described him, but his importance was magnified by his presence on the Burke and Wills Expedition, an exploration that plays a critical role in how Australians view their desert landscape, and how the aerial imagination treated that surface during the twentieth century. Bonyhady, in fact, curated the exhibition and wrote the accompanying catalog for *Burke and Wills: From Melbourne to Myth*, which traces how everyone from Becker through Sidney Nolan and other painters of the twentieth century used that journey as inspiration for their work. One body of Mandy's work is a notable addition to that canon.

Becker, a German scientific illustrator born in 1808 who arrived in Australia in 1851, traveled with Guérard, took meteorological observations for the German geophysicist Georg Neumayer, and became a leader both in Melbourne's artistic and scientific circles. Like Neumayer, Becker was another devotee of Humboldt. While the scientist was encouraged by Humboldt to establish a geomagnetic observatory in Australia, the artist accepted the famous man's dictum that artists should faithfully depict nature as specifically manifested in local environments, not as idealized settings. But Becker was also steeped in the Romantic painting tradition of Caspar David Friedrich, and his interests tended toward the more spectacular examples of fossils and phenomena. Thus, while he was a meticulous natural history draftsman, he tended at times to make statements about geological and a biological matters that were a bit speculative.

During the late 1850s Melbourne was flush with money from the gold mines, and its citizens funded an expedition that would be the first to cross from the southern to northern edges of the continent via the unexplored central deserts. Becker obtained a place on what became formally known as the Victorian Exploring Expedition of 1860–1861, which Neumayer helped organize. The journey was led by former military officer Robert O'Hara Burke and William Wills, the expedition's surveyor and astronomer who became Burke's second-in-command along the way. Seven men died during the journey and only one man made it back to Melbourne alive after having completed the round-trip. Among those who perished were Burke and Wills, as well as Becker, who suffered from dysentery and scurvy.

Becker left behind more than seventy expedition sketches and paintings, many accompanied by notes commenting on the local geology, fossils, and the effects of floods and other events. His most famous desert painting, a 1861 watercolor titled *Border of the Mud-desert near Desolation Camp*, shows two dingoes standing at the edge of a desert playa near tracks that disappear across it. On the horizon a mirage beckons, a silvery blue band that offers relief from the hot oranges and yellows of the desert. A line of riders on camels emerges from the mirage in the same cool hues, thus confounding the viewer's sense of reality—there is no way to tell if they are real or a projection of the imagination.

Just as the romantic sublime was a governing aesthetic for Americans viewing the West during the nineteenth century, so it was for the Europeans settling Australia. The scenic climaxes in both countries were defined by and represented through paintings based on European landscape traditions, confirmed in early photographs, and codified by governments into parks with official vantage points complete with interpretive plaques and guidebooks. Contemporary landscape painters, at least the more intelligent ones, know the history of their genre and how to mine it for technique while wrestling with its conventions. You just can't do the sublime anymore without acknowledging the effect of humans on the landscape. That explains exactly why Mandy, in dealing with some of the same landscapes over which Burke and Wills wandered, includes industrial structures, such as mines on the horizon line, that now exist in sites visited by the expedition, versus a romantically empty desertscape.

A tall woman with dark curly hair, Mandy Martin was born 1952 in Adelaide and raised in South Australia, which accounts for an accent that takes some getting used to. Her father was a biologist who organized expeditions collecting Gondwana marsupials, plant species, and fossils not only around Australia but South America, California, and New Guinea. He took her along on field trips in the Australian deserts and for a stint in California and the Sierra Nevada. When he passed away in his early seventies he was still actively working on the genetics of *nothofagus*, the southern beech that apparently was a Gondwana tree. Mandy studied art at the South Australian School of Art, and quite logically ended up teaching at the fledgling Environmental Studio within the art department at the Australia National University in Canberra.

Mandy has long been organizing her own expeditions into the interior to ranchlands, the desert, archeological and industrial sites, and some-times retracing the classic expedition trails. More often than not her real focus is areas of environmental sensitivity. It was on one of those trips in 1995 that she invited Guy along to observe and write about the ecology of northwestern New South Wales. Guy's mother was a well-known Aus-tralian children's author who was the first one to introduce Aborigines into that literature, and he was just beginning work on his dissertation about "The ethnographic relationship between the social system and the ecosystem in the Rangelands of Australia." The rest, as they say, is history. Or perhaps more accurately, art history.

Mandy's characteristic landscapes with their site-specific pigments—a practice Becker also employed, albeit from necessity—are layered with gesture and meaning. A fine example is her series *Reconstructed Narrative, Strzelecki Desert: Homage to Ludwig Becker*, which she painted from her 1991 trip to the Moomba Gas Fields. As Roslynn Haynes points out in *Seeking the Centre*, a book exploring the cultural depiction of the Aus-tralian Desert, Mandy's exemplary *The Effect of Refraction* in the series replaces the mirage of water with the cracking towers of a refinery, equat-ing oil with water (and Becker's camels) as a Western necessity of life. Scrawled in the paint at the bottom of the canvas are her notes on local pigments and landscape features, compass directions and latitude, and text lifted from Becker's journal. It's gratifying to see an artist and a woman resurrect the role of the exploration leader in Australian culture, which like the American West had long been perceived as the province solely of males bent on conquest. Wisely, and unlike the sometimes overly ambitious Becker, Mandy has not taken on the role of scientist, but secures the company of them.

Mandy is also one of the few Australian artists to have crossed the deep cultural divide between Europeans and Aboriginals. She has worked along the Lachlan River with her neighbor, the Wiradjuri artist Trisha "Trish" Carroll. Mandy painted at (but did not directly represent) carefully selected local Aboriginal sites where it was permissible to do so, canvases upon which Trish then added the appropriate gestures to indicate the presence of Aboriginal Dreamings. The sanctity of the sites was preserved in what scholar and critic Sasha Grishin postulates as a postcolonial collaboration, a reconciliation between the European and Aboriginal systems of image making. Mandy's ochre and black landscapes with Trish's waving lines laid over them like a psychic map—I'm thinking in particular of the large five-panel panorama *Absence and presence* from 2006 that Mandy showed me in her studio—are as seductive as a view by Rosa, yet unsettling in ways he never could have imagined.

Among many of Mandy's attributes are the network of friends and colleagues into which she has inducted me, and her willingness to arrange trips in the air, on the ground, and even underground. She is one of that cadre of people convinced that, if I'm writing a book about how we see the world from above, then I should also write one about how we see it from below. She's determined to get me into an underground gold mine somewhere, as well as into caves that Trish knows about. But aerial matters come first, so after a light lunch we drove over to George's place.

The airstrip, what here is still called an "aerodrome," consists of a grass strip 650 yards long, with its metal shed hangar toward one end and a windsock at the other. George flies a 1980 Cessna 182, one of the most popular aircraft in the world, a four-seater built of aluminum with a 230 horsepower engine. He dragged the plane out of the hangar by its tail, and we clambered in, buckled up, held cameras at ready, and took off for a tour around the countryside. The airstrip is 2,350 feet above sea level, and with three of us and a full fuel load, we managed to clear the power lines just past the end of the airstrip by only about 50 feet. "Pretty high elevation here without a turbo," George commented. I kept my thoughts to myself.

Mandy had asked George to follow the local watercourses as they flow into the Murrumbidgee, a river that then feeds the Lachlan, which eventually joins the Murray. I had asked him to comment on the environment as he saw fit, and he began before he'd even leveled off the plane. He pointed out the window with his chin. "Roads tend to follow the hard stony ridges, but flying you can see the soft country in between and see what the ground coverage is like." One thing that was immediately obvious was that where trees have been left in the ground, there was less erosion and fewer cutbank gullies. The other thing I ascertained was that there were livestock ponds everywhere, their surfaces glinting in the afternoon sun as we flew northwest toward the Cadia mine. And many of them were dry.

Our first objective was the pit, and it took him only a few minutes to put us over the operation. Cadia is your standard gold and copper operation, a gaping orange hole excavated about halfway down to the 1,675-foot mark it will reach when finished. Two enormous tailing ponds, each about a square mile in extent, stood nearby. What was amazing to me was the amount of greenery surrounding it. Unlike Bingham in Utah, a much larger operation, the environment here wasn't a moonscape. George, in blatant disregard of what are rumored to be rules about overflying the mine, circled over everything twice. Mandy was fascinated, in particular by the tailing piles, which she has painted before from ground level, and the edges of which are clearly visible through binoculars from Pennyroyal. She's been thinking about doing an aerial view of it, and the camera clicked away (Plate 8).

The Cadia deposits of gold and copper were created when volcanoes were active here before the continent was even part of Gondwana. Gold was first discovered here in 1851 and has been mined off and on since 1870. It wasn't until the multinationals got involved in the 1990s that the economics were feasible for industrial mining, which required that millions of tons of overburden first had to be removed. Over the projected thirteen-year lifespan of the pit the company will dig out 200 million tons of ore and 286 million tons of waste material. About 73 percent of the ore will be gold, the remaining 17 percent copper. Each of the trucks we saw below us climbing up the three miles to the crushers carried about 250 tons. That's lot of trips in only thirteen years.

Once the ore has been crushed, it's processed with water. The ore-bearing slurry is sent through a pipe about a yard in diameter to a nearby town where it's prepared for shipping to Japan and smelting. The waste-bearing slurry is sent to the tailings ponds, each of which will be about three hundred feet deep when full. As the material settles, the water is recycled. Our aerial vantage point made apparent two opposing aspects of the mine. One, from observing the amount of water held in a reservoir above the dam, as well as what was flowing into the tailings ponds, it was obvious that a lot of water was being used. But two, the land was relatively healthy compared to that around the Bingham operation in Utah because the smelting—and attendant toxic chemicals—was being done in another country.

The biggest surprise, however, was the sinkhole in the middle of a pasture over which George flew us. A really big sinkhole, 150 feet deep and wider across than a football field. Cracks splitting off to the sides indicated more collapse was soon due. As it turns out, the Cadia pit produces less than half of the gold; the other portion is mined underground through a tunnel dug from one side of the pit into an ore body, named Ridgeway, which lies 1,650 feet beneath the surface of the grassy field. At the end of the tunnel is a bulb of gold in sheets of vertical quartz the bottom of which has yet to be discovered.

Most mineral deposits in Australia are found using airborne electro-magnetic (AEM) technology, through which responses to electrical pulses from an aircraft can detect and map ore bodies almost a thousand feet below the surface. Ridgeway was deeper but enough aeromagnetic anomalies were detected to justify drilling the deep holes needed to confirm the findings. Production started in 2000. Once miners reached the ore, they excavated a chamber into it, blew the ceiling down, hauled out the material, and repeated the process. The tunnel actually runs under a working farm, and one outbuilding has already been swallowed. It's one of the riskier ways to mine gold, this semi-controlled collapse, but the ore at Ridgeway is hugely profitable.

Mandy had already arranged for us to tour the mine by foot the next day, so we left it behind to head over to Wyangala Dam where we could assess the effects of the drought along the Lachlan River. We were flying at four thousand feet and passed a Wedge-Tailed Eagle with its six-foot wingspan. I'd never seen a raptor flying alongside before, and I had a proprioceptive twinge, as if we were a big bird, too, and could feel the air through muscle and tendon. That's been happening more and more as I fly, especially in the smaller aircraft, and it occurred to me there was another cognitive issue there to investigate.

George took us up into the hills so we could follow one of the tributaries feeding the reservoir down to the dam, and we flew over a smaller dam built upstream that has long been silted up. You hear about the inevitability of dams in the American Southwest along the Colorado River filling up, but I'd never before seen a dam turned into a waterfall. "You only get to build a dam once—it's all about the catchment, the big picture." George's implication was that the graziers who built the dam weren't thinking about the erosive damage they were causing upstream by cutting down gum trees and letting livestock trample the streambanks. The farther west we flew, the more the trees looked less like "full-topped broccoli," George's rough measure of health, and more like sticks with leaves. "Lots of erosion here," he added, "The land's not holding together well." We reached the top of the reservoir and could see the top of the dam several miles downstream. Our pilot's observation? "Most dams in Australia look like football stadiums now because they're empty."

Construction on the original Wyangala Dam was started in 1928, a handsomely curved concrete structure, but in 1971 an enormous new rock embankment was built immediately in front of it in order to raise the level of the reservoir. The new structure, much larger and longer than the previous dam, was an aesthetic disaster, but it raised the reservoir seventy-seven feet, increasing its storage capacity 400 percent. The newer dam tops out at the height of a twenty-five-story building; when the reservoir is full now, it holds more than twice as much water as Sydney Harbor. Ah, but that's the rub—when full. As we flew over it, the reservoir was down to only 3.6 percent of capacity and the old dam was almost completely exposed. Most of the reservoir was a mud flat.

As we followed the Lachlan downstream we dropped off the edge of the tablelands and into the Lachlan Valley proper, an ancient alluvial plain over which the river meanders. Even when the river is high it's not exactly energetic because the land is so flat. Its course was formed when the waters were in flood and left behind a series of ponds where silts were dropped. Between the successive silting events and the lowering rate of flow, the land gets less rich the farther it is from the mountains.

The Lachlan we were flying over was still very fertile, however. Vine-yards, big houses, a golf course in Cowra. George took us down to three hundred feet above the fields so we could observe the well-irrigated crops, which put Mandy's teeth on edge. Pennyroyal is reliant on rainfall for water. A huge flock of sulphur-crested cockatoos rose up over the river, the large white birds with their yellow crests an almost tropical sight. George noted that, as we continued west, the land was drying out.

"We need to change from annual to perennial plant cover, and stop peeling the skin of the land every year. Loss of biodiversity causes deser-tification, and we're hurrying along that course as fast as we can. We can't turn it back, we're not all going to leave, but we have to create something new that's old."

At 4:00 P.M. the low sun glinted ahead of us on dozens of stock ponds. "Go ten miles to the right or left, it would look the same. It's amazing how much water we're storing in Australia." It reminded me of flying over Kansas with Terry Evans, the same flat land riddled with human-made ponds. Then we were looking at wheat silos, the river was getting smaller, the land redder, and incipient dune formation was visible. "All that good silt was dropped earlier," George remarked as we reached Forbes, our turnaround point. From there to Alice Springs it's pretty much flat, red, dry. George shook his head. "The great Dividing Range and the Great Desert are going to meet if this keeps up."

To our west, as the country continued to drop, was the sheep grazing country along the Darling where the photographer Ruby Davies grew up riding with the Aboriginal stockmen who ran her father's sheep. She went to school in Sydney, where she still lives and teaches, but goes home frequently to family property. In 1991 she saw the river drying up near her brother's property for the first time. Algae blooms fouled the riverbed, and the customary kangaroos and birds were absent. She began to wonder what was going on upstream, and what she found surprised her. "We didn't have any idea about the extent of the deforestation, or the amount of cotton being planted. People were sort of excusing the condition of the Darling drying out, saying it always had on a seasonal basis— but not like this."

Being a photographer, her reaction was to document the situation. She'd been using a standard 35 mm camera for her work in Sydney and other cities, but needed something that worked on a slower timescale, that would correspondingly bring her to a more contemplative state in line with the country. So she built a wooden 4 x 5 inch pinhole camera, taking long exposures without using a lens, but with a focus that extended from the front of the camera to the infinite. She could picture the manmade furrows and other patterns from the ground, but she wanted to show how the agricultural grid of the irrigators cut across the natural patterns created by the intersection of contour lines with flowing water. Her father had introduced the use of aircraft for mustering livestock in the region during the 1940s, and one of her brothers was also a pilot, so she knew those patterns were apparent from the air. She was also an admirer of the work of Terry Evans, so she took to the air.

The first thing Ruby realized was that, yes, that intersection of the natural and the unnatural was visible. The second thing was that she wasn't flying the plane, and as a result she couldn't get the images she wanted. In 2002 Peter Murphy, a panoramic landscape photographer, suggested that she try kites. She went online, looked up the "KAP" (Kite Aerial Photography) site run by Charles C. Benton, a professor of Architecture at the University of California at Berkeley, and started building her own rigs. Most kite photographers, she found, use disposable cameras because they're light. She needed better resolution, however, so ended up lofting a small Zeiss $2^{1/4}$" Bellows camera that was still small and light, but that offered larger negatives with more detail. Its only drawback was that she had to bring the kite down after every shot and rewind and remeter. She doesn't get a lot of images and it's very physical.

"You have to previsualize, which I can do because I know the land— from both ground and air—I know all about the light angles and shadows.

I tried at first photographing at midday to avoid being romantic, but the results didn't look good, so I went back to low-light angles." The process remains almost comical in its intensity. She parks her car, runs out a thousand feet of kite string attached to the bull bar on the front, then trots back to the car. She flies a Japanese Rokaku used for long-line fishing. "You get those kites set right, hooked into the air, and they just sit in the wind. It can be at a high angle or low, depending on wind speed and such. I have a separate string that runs up to the camera to trip the shutter."

It can take Ruby up to two hours just to get the kite in position. After she takes a shot she pulls the kite down to about a hundred feet, which brings the camera within reach. She rewinds, remeters, and the kite goes up again—if she's lucky. She flies it about five hundred feet up in the air. With a grin she told me "Sometimes you can even drive the car with the kite attached!" All this work and she's gotten perhaps ten successful aerial photographs. In her master's dissertation done about the Darling project, she notes that the images often resemble Aboriginal dot paintings, the red earth patterned by vegetation and scored by the tracks of sheep.

As George banked the plane, I looked west out over the dry earth and believed George when he said the desert could come this far to meet the mountains. On our route back he detoured us a bit to the north in order to fly over the north shoulder of Mount MacQuarie and the wind farm there, the first large-scale such project in the country. Fourteen turbines standing almost 150 feet tall with 150-foot-diameter blades rotating slowly next to the Cessna provided another haptic moment, although not as happy a one when flying next to the eagle. This had more to do with slicing-and-dicing, just as it had when I was flying next to the wind farm in the Berkshires with Joe Thompson. As we circled the peak George chimed in. "Not a hill between here and South Africa taller than this one—four thousand feet." Then we were dropping fast toward the ground, the mouth of a narrow gorge approaching fast . . . .


The next day, our feet firmly on the ground, Mandy and I head over to the Cadia mine to take an afternoon tour. After the mandatory sign in, orientation and safety drill, we're issued a magnetic swipe card in exchange for our driver's licenses to get us through the gate. The mine has its own explosives batch plant on site, so security is a bit more elaborate post-9/11. Mandy's led several field trips here with students from the Environmental Studio at the Australian National University (ANU) in Canberra, and is already on speaking terms with Nedra Burns, the community liaison officer. Nedra promptly loads us into a pickup truck, first for a peek over the edge of the pit and then out to the tailing ponds. In addition to the original Cadia pit, which will bottom out in 2013, and the underground Ridgeway operation, Newcrest, the company that owns the mine, is proposing to open Cadia East, an even larger adjacent mine that will also be both a pit and below ground mine. It's projected to have a lifespan of twenty or thirty years. As Nedra points out the location of the new project, water trucks rumble beneath us in the pit, spraying water on the road. It might be a drought, but dust abatement is mandated by federal air quality standards. When I ask her about it, she adds: "But more to the point, it's about local landowners." By that she includes Mandy and Guy, among others, who don't need to have their fields smothered in toxic dust.

Pine forests have been planted all around us. "Yours?" I ask. She shakes her head. "That's Department of Forestry, not us. We wouldn't do that because of fire hazard and they're not indigenous. But all those trees are already sold, so they have an investment." Newcrest, as it turns out, plants only native flora in the buffer zone surround the property, including between ten and twenty thousand trees a year.

We drive over to the topmost tailing pond and walk along its edge. Roo tracks are everywhere. A large pipe, almost more a black hose, runs around the perimeter of the pond, and "spigots" every few hundred yards lead down into the mud. They alternate pumping waste slurry into the lower and upper ponds in order to keep each one wet, otherwise they dry out and dust would blow all over the valley. The pond is already 295 feet deep at its lowest point and will be full at around 310 feet. The top of a dead gum tree sticks up out of mud near the shore. "We don't pull out the trees, as if we did, the tailings would leach down into the root system and into the water table," Nedra explains.

Nedra points out how they are beginning to contour the older overbur-den piles, "at least in imitation of what the hills look like." They've been reseeding them from the air, which is a bit tough during dry times, but it seems to be working because native grasses have some drought tolerance. The idea is to rehabilitate the land back to grazing, although with the development pressures coming, I wonder if Newcrest won't end up turn-ing it into subdivisions, as is happening in Utah with Kennecott's Day-break community.

Nedra has a meeting to attend, so she turns us over to Alex Reed, a twenty-nine-year-old environmental scientist who got her degree in "soils at uni." She takes us up along the Cadiangullong Creek diversion, said creek once flowing directly over where the pit is now. This is part of the headwaters of the Lachlan River, so it's being treated with great care. They've cleaned out non-indigenous plant life, but left the native ferns and deadfall. Recently they've even seen platypus in the lower reaches, the shy animals a positive and direct sign that the company is doing a good job. We drive by the first "open cut" they dug, the pit now backfilled with rock from Ridgeway, and settling. Eventually, as with the tailings ponds, they'll cover it with soil, biosolids, and plants. They'll do the same with the big pit if Cadia East goes through, fill it up with waste material and rehab it. We pause to look down on a dam they built in 1998, which is not a small structure, and that's when it hits me how much infrastruc-ture they had to create to make this happen. They've rerouted the nature of an entire valley.

Alex puts the truck in four-wheel drive low and we head uphill to the edge of the subsidence pit caused by Ridgeway. It has a three hundred-foot buffer zone around it, then more land that's being grazed, then a cyclone fence. Vent towers for the tunnels stand in the middle of the pas-ture. Clearly they expect the pit to get much larger. Back in the truck and now in compound low we grind up one last slope to a heritage site from the nineteenth century, an old gold operation run by Cornish miners, where the landscape is eroding on the slopes right from under the mine head. It bothers Alex that she's not allowed to stabilize this with some reseeding, but by law they can't change the character of an historical prop-erty. How ironic that is, how loaded with anthropocentric bureaucratic hubris, that the company can rearrange the natural landscape, but is required to leave human ruins untouched.

It's late afternoon as we head back down to the office, the roos coming out for supper among the eucalypt. Newcrest has taken the environment here as a serious issue, and it seems more than merely a political gesture, or even commonsense for the continuing viability of the business. Nedra and Alex are obviously aware of the fact that mining causes havoc with the landscape, but are proud of how the company is conducting itself. They sponsor programs such as inviting students to plant and monitor various plants to see how they do in the revegetation program, and actually retire some woodlands to allow them to regenerate from past mining activities, as well as maintain wildlife corridors. Of all the mines I've visited, from the United States to Chile, this seems by far to be the most environmentally sensitive operation.

During 2001 and 2002 Mandy brought artists, writers, and university students to the mine during a series of five intensive working visits. The project, *Land$cape: Gold and Water*, saw notable artists such as Wendy Teakel and John Reid create new work in response, but also had scientist John Chappell to make sculptures about the mining process. Of particular interest to me was painter Naomi Greschke, who created aerial views of nearby industrial agriculture. Mandy has noted that the monoculture of corporate farming creates a landscape devoid of scenic amenities, just as does the mine. From literally a different perspective, Greschke wrote about how from several thousand feet up "The ugly, barren or absurd can take on a new aesthetic value. The eye naturally selects harmonious elements of the larger picture. Organic shapes or lines can be found in the unnatural."

Mandy herself created one hundred small paintings about the mine titled *Not in arCadia ego*. She used a gold palette incorporating river sand for her pictures of the Belubula River, and a copper palette with tailings and sulfide concentrate from the mine for the Cadia paintings. The paintings are arrayed around a large canvas diptych of the tailings dam with a text by Chappell inscribed in the paint that points out less than a gram of gold is taken out of a tonne of rock (about 1.1 tons for Americans). The catalog includes several essays, including one by Guy where he tells how gold mining, cattle raising, and the growing of water-thirsty cotton and grapes in the region all take a toll on the environment. Where Mandy emphasizes that short-term economic gain from gold ore may be less valuable than the long-term value of undisturbed land and water, Guy urges for all the land users to concentrate not on their differences, but commonalities that can be enlarged. That's a viewpoint that to me lends itself to an aerial perspective.

Landscape is what a community makes out of land, is how I would put it, and the human community is inseparable from the environment, from nature. Most Australians live on the coastline, yet their environment, hence the future of their communities, is shaped profoundly by the interior, the majority of the continental landmass. Understanding the desert heart of the country is essential. While there is a significant body of aerial photography in Australia, I'm coming to believe that it's as much through aerial paintings that the landscape of the interior has been shaped in the minds of Australians. My next trip, then, is to drive through that interior, fly above its lowest point, and consider those images, the great desert canvases of Sidney Nolan, Fred Williams, John Olsen—and Mandy Martin.

# 14. Lake Eyre:
## 49 Feet Below Sea Level

*In the midst of his dream, a person who is flying declares himself the author of his flight. A clear awareness of being able to fly develops in the dreamer's soul.*

*I will, therefore, postulate as a principle that in the dream world we do not fly because we have wings; rather, we think we have wings because we have flown. Wings are a consequence. The principle of oneiric flight goes deeper. Dynamic aerial imagination must rediscover this principle.*

—Gaston Bachelard

**The town of Marree this noon in late May is tolerable, unlike in** January when the temperature here can reach almost 120°F. Even so, its sparse collection of buildings laid out on a grid is dusty and pale as if heat-soaked. Kim Mahood and I are about 430 miles north and inland of Adelaide, and within an hour's drive of Lake Eyre. The playa is so large—about 3,670 square miles—that early explorers thought there was no way around what looked to be a permanently dry basin. Two of Australia's most legendary dirt roads now skirt the basin. The Birdsville Track heads northeast across the Simpson Desert and into Queensland, and the Oodnadatta, which Kim and I have been following, bears northwest to Alice

Springs. Between the two sits Lake Eyre, and Marree is the point of the rounded V formed by the two routes.

Marree, known originally to Euro-Australians as Hergott Springs, was the launching-off spot for the 1885–1886 camel expedition led by surveyor David Lindsay, an exploration to the Gulf of Carpenteria on the north coast of the continent. Lindsay wandered briefly into the Simpson Desert, but more importantly, was the first leader to include an official photographer on an expedition in Australia. The town is also where C. T. Madigan finished his 1939 trek across the Simpson Desert, which holds the world's largest field of parallel sand dunes. These are also the longest such dunes in the world, and the tallest one—the legendary Big Red—reaches a height of 130 feet. Madigan's journey was the last exploration of a major desert in the country. At its heyday around the beginning of the twentieth century, Marree held perhaps as many as six hundred people. The population now hovers between sixty and seventy, and the main source of business is tourism. People drive here to see Lake Eyre and take the Birdsville Track, which crosses Big Red and is considered one of the world's classic motoring adventures.

Between us and Lake Eyre, our destination for this portion of the trip, is the Plain of Illusion, named for the mirages that frequent this driest part of the continent. Marree was, logically enough, originally settled by camel drovers, the animals imported from India at the suggestion of Becker and others who knew they would function well in the climate here. Indeed, they became the principal mode of transportation in the Lake Eyre Basin during the nineteenth century until the railroad arrived in 1881. Today the herds of feral dromedaries are so large, totaling as many as three-quarters of a million animals, that they're a pest and are exported to Saudi Arabia.

Kim Mahood, a friend of Mandy's, is another of Australia's more notable artists and writers. She's a slender and intense woman with shortly cropped dark hair. Born in 1953 in Perth, her name and coloring lead many people to assume she's descended from the Afghan camel drovers who settled places such as Marree, but she's actually of Irish descent. Kim grew up in the central desert, her father a grazier and her mother a well-known writer. Over the last three days she's driven the two of us nearly 1,200 miles from Canberra, much of it over dirt roads. We're traveling in that ubiquitous vehicle of the outback, a virtually indestruc-tible 1997 Toyota Hilux. Basically—and I do mean basic—it's an extended cab pickup with a wide aluminum tray on the back, versus the typical American pickup box. Much of the truck's 108,000 miles has been on desert roads like the one's we've been following, and red dust fills every crevice. Last night we'd camped at Farina, an abandoned farming town settled in 1878. The town was laid out on an expansive grid meant to accommodate hundreds of settlers on the theory that rain follows the plow—plant wheat (hence farina) and the clouds will come. That was an idea popular during the Land Rush of the American West, where it had a similar rate of success. Kim remembers coming through here on the rail-road as a kid, but now the two hotels are in ruins. The nearby cemetery has graves with headstones in both English and Arabic for the camel driv-ers who died here, their inscriptions resolutely faced toward Mecca.

Before driving north to Marree this morning, Kim had put more air in the tires from a compressor she runs off the engine. The problem here isn't so much sand, which would require her to lower pressure and fatten the tires for more flotation; it's rocks, and by making the tires harder and the treads wider she lowers the chances of them picking up sharp stones. The Hilux is a standard four-wheel drive utility vehicle for rural stations, and Guy runs two of them out at Pennyroyal. Kim has modified hers consid-erably, adding a canvas top over a metal frame and a capacious custom-built wooden campbox the front of which folds down as a kitchen counter. The back is filled with gear for her four-month journey westward into the Tanami Desert to work on a mapping project with Aboriginal people. How people pack their vehicles tells you everything. This one has its priorities clearly established by where things are, and the fact that everything fits and is easily found and repacked. The foot-high sides of the tray drop down for access from three sides. Food, tools, water, jerry cans of diesel, plastic drawers of spices and condiments. A tent, kitchen gear, and two swags for sleeping. I borrowed mine from our friend John Carty, a young anthropologist working on paintings produced by Aborig-inal artists in Balgo, one of Kim's destinations. The truck brims with gear.

Camping in Australia is a very odd experience for someone used to down-filled nylon sleeping bags, and the swag tells it all. It's an archaic piece of outdoor camping gear every Australian devotee of the bush is obliged to own, a waxed canvas bag with a foam pad and woolen blankets sewn in. It's incredibly heavy and cumbersome; you couldn't take one backpacking if your life depended on it. But it's exactly right for throwing on a ground carpeted with burrs and thorns, which is typical of our better camping spots. John has added a synthetic blanket picturing African megafauna and a pillowcase of cotton on one side and sateen on the other. It's the height of luxury and very, very comfortable.

When we drove out of Canberra, we headed west along the Murrimbidgee to its confluence with the Lachlan River, and followed them to Mildura, which is where the Darling joins in. Turning north we made it to the mining town of Broken Hill, then cut left to the western edge of the Murray-Darling Basin, which we followed up to the Lake Frome playa. My route from Guy and Mandy's down to Canberra, then west and north with Kim to Lake Frome made a giant U around the bottom of the basin.

From the dry lake we once again turned west and slipped through the small Italowi Gorge in the Flinders Range to reach the shores of the larger but equally dry Lake Torrens. From there it was virtually a straight shot to Marree and the Lake Eyre Basin, one of the world's largest inward draining watersheds with no exit to the sea. Our dirt road had followed the 1870 telegraph line, as well as the old Ghan (shortened from "Afghan") railway route that links the southern and northern coasts of the continent, which had been the dream of Burke and Wills. Lines that cross the continent with any kind of common sense get reused over time. The Oodnadatta Track, for example, was originally an Aboriginal trade route that followed a series of springs. The telegraph and railway followed, then the road to Alice, and today it's a popular route for tourists in their Land Rovers and other 4WD vehicles.

Kim has parked the truck outside the Oasis Cafe to fill up with petrol, let her boisterous young dog Pirate out for a run, and to hand roll a cigarette, which she sticks in one corner of her mouth while contemplating a young guy in slacks and a shirt with epaulets who is writing on a blackboard. Something about being able to charter a plane for a flight over Lake Eyre. Her eyes narrow as she looks at him, then me. Yup, it's gotta be done. His name is Ben Newton and he works for Central Air Services.

We arrange to meet him at the aerodrome at 10:00 the next morning. After I consume an appallingly large hamburger upon which fried eggs have been placed, a combo referred to continent-wide as "the works," Kim and Pirate and I climb back into the truck and head across the red gibber Plain of Illusion.

The gibber—Australian for that mosaic of tightly packed stones one layer deep known as desert pavement—is flat, red, seemingly unending, and so reminiscent of images sent back from Mars that I ask Kim to stop so I can lie on my belly and take a photo. About halfway to the lake we cross the Dog Fence, which runs 3,555 miles across the country. Like its more famous cousin, the shorter rabbit-proof fence in the far western part of the country, the Dog Fence was built as an exclosure to keep an animal out of one part of the country, in this case the wild dingo from the sheep to the south. Dingoes are thought to have descended from the wolves of southeast Asia, can grow up to more than fifty pounds, and are agile enough to climb trees. An individual dingo can kill up to six sheep a day. Needless to say, the fence is by necessity a formidable defense. It's six feet tall, extends a foot underground, and the land is cleared back from both sides and patrolled by maintenance crews. Both traps and poison baits are set out along it, and when I get out to open and close the gate, Kim is careful to keep Pirate in the truck. The fence is the longest continuous human-made structure on earth. To see it run across the gibber plain with its mirages of water is to feel like you're in a very strange story, indeed.

We wind our way through white sand dunes along the Goyder Channel that connects the relatively small Lake Eyre South and end up parked above the shoreline of Price Peninsula, which extends out into a lobe of the playa, the Madigan Gulf. Lake Eyre is so large that I can't even see the main playa from my elevated viewpoint several yards above the flat. I wonder briefly about access to the playa for the truck. First, that's prohibited by law, but second, the tracks of those who have attempted it tell an interesting story. Tire prints head out a few hundred yards and get progressively deeper as the vehicles break through the crust and start to sink. And then they stop. That's it. They must have backed out over the same tracks. That, or been lifted out by UFOs with tractor beams. In any case, we're not tempted.

Lake Eyre sits in a fault basin, a shallow bowl created by movement in the earth along fault lines that create a depression on the surface. The basin has been covered off and on with water for 25 million years, but the depression as it is today was shaped about 120,000 years ago, and it wasn't until 18,000 years ago during the last glacial period that the modern-day lake formed. It does flood every few years, as it is apparently doing now with runoff from recent rains to the north. We can't see it from here, but hope to spot the waters from the plane tomorrow. And the lake has actually filled up completely three times during the last 150 years, making it by far, if temporarily, the largest lake in Australia.

I go for a short stroll on the sand dunes and find the tracks of rabbits and lizards, and a hopping nocturnal marsupial. The tracks of a barefoot person, a dog, and my shoe prints. That's it for what is the most easily accessed point of Lake Eyre. Even in this autumnal afternoon light, the playa is so bright that I can't look at it without my sunglasses, and so large that I have no sense of its size or scale. I'm curious to see if tomorrow that will change from the airplane.

Driving out we pass a Land Rover headed in, the only other two people we've seen approaching the lake. Kim figures we'll camp at Muloorina Station, a 400,000-hectare spread halfway back to Marree and fed by artesian wells. It's not the kind of grazing property Guy and Mandy have, and able here to support only one cow per 160 acres versus the two or three cows per acre found in much of Europe and North America. It's a well-watered and pleasant spot, a small reservoir surrounded by various kinds of eucalypt, and as we set out our gear under the gum trees, I spot a Great White Egret stalking through reeds, and numerous coots and cormorants on the shoreline. Overhead in the branches of the mulga, a species of acacia found throughout the desert, are the ubiquitous galahs and corellas, two species of cockatoos widespread over Australia. And a much shyer Honeyeater of some kind is keeping an eye on us. After we get the kitchen set up and the swags out, Kim walks off for a smoke and returns with some worked stone flakes. She found them near a dry streambed, a place that would have been a logical camp spot when the Frome River flowed out of the Flinders Range through here to Lake Eyre. She brings the points back to our site to show them to me, but then puts them carefully back where she found them, making sure that no one is watching as she does so. She doesn't want anyone taking them away.

Kim is a bit of an enigma to me, perfectly friendly and open about her past, but with great reserves of experience I'll never approach because of her unique upbringing. Her father was given the post of superintendent of an Aboriginal settlement some seven hundred miles southwest of Darwin in the Central Desert. Kim's mother had studied French and literature at university. They joined Kim's dad at Hooker Creek, where he was responsible for several hundred traditional Warlpiri people recently relocated from the desert. It was here that Kim was given the skinname *Napurrurla* to indicate how she fitt into the traditional kinship structure, and which to this day places her affiliation with a particular part of the country and its people whenever she talks with Aborigines. After working on several different settlements in the Top End (the upper reaches of Northern Australia), the family relocated to Finke, a small railroad town, where Joe Mahood took a job as a stock inspector. In 1960 they moved to Alice Springs, where her mother taught French and English, and Joe continued to work as a stock inspector. Having always dreamed of being an artist, he planned to quit the job and work as a freelance cartoonist; in 1962, however, he was asked to help establish a stock route from the southern Kimberleys to Central Australia, a life-changing adventure. When offered a pastoral lease in the Tanami Desert some four hundred miles northwest of Alice, he accepted. The family spent nine years there.

The station was 1,624 square miles, more than a million acres. They could run at maximum during wet years only about eight thousand head of cattle on a property that size, but usually kept only four thousand given the dry conditions that prevailed most of the time. It was marginal country, remote and cut off from the outside world when it did rain, and everything that wasn't grown on station had to be trucked in. Kim was taught stock work by the Aborigines who worked for her dad. She was sent away to boarding school when she was older, then went on to attend university in Perth and then Townsville to study literature and anthropology, but it was difficult for her to spend even a year at each place. Sitting around the fire, the ever-present cigarette in one corner of her mouth, she admits: "I was a frustrated student and questioned everything. If I hadn't been brought up in the bush, I would have just accepted what my teachers said." Restless, she did what many young Australian artists and intellectuals do, she went to London. After two years trying to make ends meet in one of the world's most exciting but also most expensive cities, she decided she needed to return home. With no car or money for plane fare, she simply started bicycling with friend across Europe, drawing portraits to pay her way. She ended up on an Israeli kibbutz for a year, where her cattle station talents were discovered, and she was put in charge of growing corn. She finally made it back to Australia in the early 1980s when her mom arranged for one of her paintings to be sold in Sydney and she could afford a ticket home from Athens.

Deciding to take her painting seriously, she studied painting and drawing at the Julian Ashton Art School in Sydney during the first part of the decade. The school, founded by an artist in 1890, is the oldest arts education establishment in the country and remains a bastion of traditional technique in life drawing and landscape. Among its many notable students was one of the major figures in Australian painting, John Olsen. During the same time, she studied at the Tom Bass Sculpture School, likewise a purveyor of mainstream aesthetics. 1993–1994 found her living in Brisbane and not far from her family's latest property, and she soon found work teaching in the Flying Art School, a project that sent artists around the country. She worked around Queensland with most of her students for the one-and two-day workshops with local station women. "Because I was from the bush, I could challenge my students. They wanted to represent their place. If I gave them that first, I could then give them something more abstract, more conceptual. You have to allow people to see in a way they're comfortable with, then to see what the light does, to paint tonally." Tone and color are constant references made by Kim as we travel, and the first time she came back to Alice after an absence of sixteen years, it was the light that struck her, that clear hard light.

My ten days with Kim from Canberra to Alice is just the start for her of a four-month road trip. After we reach Alice and I return to Canberra, she will go on to what's now known as Tanami Downs (formerly the Mongrel Downs station her dad ran). She'll meet with the Warlpiri, the traditional owners there, to work for two weeks on the map they are creating from stories and local knowledge. This project started out as a way of preserving place names, life histories, and sometimes traditional stories, but is increasingly conflated with land claims. Gold was first discovered in the Tanami in 1900, but it wasn't until the 1980s and the advent of cyanide leaching that it became profitable, and royalties from the mines now provide a substantial income to the Aboriginal people who can prove traditional claims to the country where the gold is located. Kim's not exactly comfortable with the politics, but is committed to the project, and is careful to make it clear that the map is to record cultural knowledge and that it has no legal or authentic Aboriginal art status. She fills in the basic topographical features on a large canvas laid out on the ground, such as blue lakes and streams, then everyone sits around and talks, adds features, assigns names.

When Kim is done with that project, she'll go out to the Tanami Mine with local artists to set up an "artist's camp" there for a month. They're not sure what they're going to produce, but the mine is slowly shutting down operations and should be an evocative place to make art, much as Cadia was for Mandy with her groups. Following that, she'll go into the East Kimberly region to work with traditional owners there on a map that is in progress.

Kim gave me a photocopy of an article she'd written for the Sunday edition of Melbourne's newspaper *The Age*, and it occurred to me when reading it that just as she is informed about her country by the work she's doing with the maps, so I'm being allowed a glimpse of how complicated the cultural cartography is out here.

The reality is that the country "out there" is known in ways that country encased in concrete and brick and bitumen can no longer be known. It is mapped, patterned, storied in multiple ways, known intimately and profoundly by its occupants, both black and white. There are recognized boundaries where one jurisdiction ends and another begins, ancestral boundaries, pastoral boundaries, language group boundaries, practical and provisional boundaries. They are contested, remembered, lied about, redefined, reinstated. For those who live there the knowledge and attachment to country register deep in the body, and manifest as a malaise when the boundaries are dislocated and fragmented. And I am not talking only about Aboriginal people.

To travel with people who know their country is to encounter landscape as an entity, animated, inhabited and continually renewing itself. Perhaps it is this that the traveler is unconsciously seeking.

I have proof of that last statement's validity. As we travel, Kim points out features and tells their stories, indeed making the country come alive for both of us. Kim and I have also been talking about flying in Australia, with which she has a personal relationship, as do many rural Australians. One of Joe Mahood's better-known exploits was a plane crash out on the Lake Ruth playa, which he survived with his three companions only to face a four-mile hike back to the station. It's the same playa where Kim learned to sail when it was full of water, and the site where her prize-winning book of nonfiction, *Craft for a Dry Lake*, is set. She writes in that book about the playa:

> Nine years of my family's history are layered into this lake surface, dry and dusty now, intermittently filling and drying in the years we lived here, sometimes a clear blue expanse, sometimes a shallow milky puddle, always a focus for us. They are like palimpsests, these inland lakes. Everything drains into them from the surrounding country, everything is distilled into a tracing of events and passages, erased and reinscribed and erased. This one is an important site, a red ochre place in the dreamtime. Did the tear the plane made on the skin of the lake set loose some angry spirit, and has my father's death appeased it?

When Kim told me that her father died in a helicopter accident while rounding up cattle in 1990, I couldn't help but think of the bent propeller in George King's hangar. Once flying was introduced to Australia, it became one of the more efficient ways to get around, to explore, round up livestock, deliver mail and medical services. Apart from Alaska, I can't think of a place in the world where flying is more part of the fabric of daily life. And I don't know of any place where the technology has had a more eccentric history.

William Dean managed to make a short balloon flight across Melbourne in 1858, the first lighter-than-air trip on the continent, and in 1909 George

Taylor soared aloft in the first heavier-than-air assemblage, in essence an enlarged box kite. There are photos of him taking off from the top of a sand hill on the beach in Narrabeen, just north of Sydney, and it looks more as if his helpers are actually launching a kite than an airplane. Taylor lay prone on a slot in the lower wing of the biplane and steered his craft by shifting his weight. His wife, tucking her skirts around her, piloted the glider the same day and became the first Australian woman to fly. The next year Colin Defries, another Sydney man, attempted a flight under power over a horse track, but only negotiated 125 yards of partially guided flight before crashing from a few feet above the ground. Credit for the first successful controlled flight under power in Australia is accorded the magician Harry Houdini, who was touring the continent in 1910.

The first Australian-designed and built airplane was launched four months later, seven years after the Wright brothers had flown at Kitty Hawk. If the history of aviation is a story of man's increasing speed in the world, the rate of its development in Australia reminds us that distance was very much a narrative factor at the beginning of the twentieth century. The first England-to-Australia flight didn't occur until 1919 and took twenty-eight days. The first aerial circumnavigation of the continent was finally accomplished in 1924 and took forty-four days. Flying across the continent was so daunting that it wasn't done nonstop until the legendary long-distance flyer from Brisbane, Charles Kingsford Smith, included it as a leg in the first transpacific aerial crossing. Balloons were drifting from France to Russia in 1900, but one didn't make it across Australia until 1993.

Australia did field a flying corps during World War I, which debuted during 1915 over Basra in what was then still Mesopotamia, and a civil aviation service was started in 1922. Airmail was being flown by then in parts of the country, but only on a limited basis. These records are important because it helps put into context how little Australians knew about their country from above—in terms of actually flying over it—while still relatively early in the history of the country. And it makes the 1929 accomplishments of Cecil Thomas Madigan all the more impressive.

The exploring geologist C. T. Madigan was born in South Australia in 1889; he went to England on a Rhodes scholarship in 1911, but returned home early in order to join Douglas Mawson's 1912–1913 expedition to the Antarctic. Mawson, another Australian field geologist, gave Madigan the responsibility for leading a sledging party about King George V Land along the coast of the icy continent. Two months and five hundred miles later he returned. If you want to know how tough the conditions were, Mawson's account of the expedition is titled *Home of the Blizzard*. You get the idea.

Madigan returned home in time to join the army and fight in France during World War I, then went to the Sudan as a government geologist in 1920, which is where he learned how to deploy camels in the field. I wouldn't say that he had something against the animals, but in his report of the 1929 exploration of "The Australian Sand-Ridge Deserts," he positively gloats about the superiority of the airplane over ground travel via camelback. Madigan convinced the Royal Australian Air Force to lend him an aircraft to explore the uncrossed deserts north of Lake Eyre. He notes in the first paragraph: "The greatest stories of Australian exploration are all connected to the efforts to conquer these deserts," and little quantitative information had been collected about "their menacing obstruction and endlessness." He aimed to fix that by ascertaining their nature from the air.

His initial discovery from an aerial vantage point was the consistent direction of the sand dunes, some of them up to 100 feet tall and running for as long as 130 miles. Everywhere he went over the 43,500 square miles of the desert—which he named the Simpson after the famous Australian geographer Alfred Simpson—he found the dunes ran parallel to the wind. He realized from his work in Africa that the geologist M. Aufrere had described the same longitudinal ridges in the Sahara, which the Frenchman had discovered from an examination of aerial photos. Whatever the prevailing winds were, the sand obeyed, not cresting in front and perpendicular to the wind, not building up a face and then collapsing as a transverse dune, but eroding at one end and building up at the far end in parallel with the wind. And if there were no meteorological records for a region, by examining the dunes you could ascertain the primary wind direction. This was also true, as he later observed from aerial photos taken by Mackay and others in the 1930s, everywhere over the 235,000 square miles of the Great Victoria, Great Sandy, and Gibson deserts. The sand ridges that covered nearly 10 percent of the continent were extraordinarily consistent, and were the greatest system of such dunes in the world.

Madigan based his theory of how such dunes formed—from small beginnings as the wind makes little runnels in the surface—by having watched the same process in the Antarctic. He related them to the *sastrugi* found there, named after the Arctic features of the same kind, and noted that they were useful navigational aids. When I was in the Arctic working with the Inuit as scouts for NASA, I was taught the same thing. Even in a whiteout with the wind blowing from seemingly random directions, let your feet find the *sastrugi* and you will know your direction.

The geologist knew that the theory required there to be a hard floor beneath the sand, which he verified from reports of other explorers, and then himself in 1939, when he became the first scientist to cross the Simpson on the ground, a voyage that took twenty-five days with camels, versus the hours in which he had traversed the same terrain airborne a decade earlier. As I noted earlier, he ended the trip at Marree, and there were still camels bringing supplies to the town until 1956. That was the year trucks took over the job.

After dinner, our fire out for the night, Kim settled in on the other side of the truck with Pirate, and I lay on my back in the swag to contemplate the sky and think about flying over Lake Eyre. Images of flying over the Australian desert have been published in the popular press here since at least 1934, when the magazine *Walkabout* started publication and began using them to promote the country as an international tourism destination. Despite the role that aerial photography has played in romanticizing the desert, I find that the relationship Australians have with their land from the air has best been expressed by its painters, unlike in America where it's been more the province of photographers. Exploring the interior of Australia has mostly been conceived, promoted, and consumed in a variety of media as a romantic quest, and flying is no exception. Painting, because it is popularly considered a more subjective medium than photography, lends itself more to romanticism, and never more so than with depictions of geographical extremes, such as the lowest point in the continent. And that brings me to a story people here keep telling me, perhaps apocryphal, perhaps true.

This part I've confirmed in news reports. There was a pilot who wanted to see if he could fly below sea level, so he went out over Lake Eyre and dipped as low to the ground as his altimeter told him was safe. He did, indeed, get below sea level, but the altimeter couldn't register the fact, so he crashed. The wings, fuselage, and engine were in good enough shape to fly, but he had smashed his propeller. Too far from the shoreline to survive walking out, as had Kim Mahood's dad, he had to find a way to fly. Finding a piece of driftwood brought onto the lake by a flood at some point, he carved it into a new propeller, mounted it, and was able to fly out.

It's a nice story, although not one I can corroborate, but the one about Kurt Johannsen flying out of Alice Springs in a Tiger Moth in 1950 is a well-known Australian escapade. Looking once again for that fabled gold deposit at Lassiter's Reef, he and a companion took off headed west by dead reckoning and no radio. They landed on a playa for the night, but when attempting to take off the next morning, the wheels broke through the surface of the clay pan, the nose dived down, and the prop lost about a foot on both ends. First Johannsen and Jimmy Price used the busted prop as a shovel to dig out the plane. Johannsen then apparently considered trying to carve a new one from a branch off a nearby desert oak, another species of acacia that thrives in the desert. Instead, he ended up cutting off the splintered parts of the prop, which left him with about half its former length. He obtained the best trim he could by balancing it on a screwdriver and whittling away until it was symmetrical.

Johannsen left Price with a primitive but workable water still he'd devised out of two jerry cans, and from which fresh water could be distilled from the brine under the clay. Then he revved up the engine as high as he dared and tried to take off. No luck. He waited for a breeze and that did the trick. The propeller vibrated badly, however, and he could only get up to tree height, not enough altitude to allow him to obtain a bearing on where he was. Then he spotted an eagle riding a thermal and he jockeyed over to the updraft. Johannsen made it the 125 miles to the nearest community with about fifteen ounces of fuel left, and made it back six days later to rescue his partner, who was doing fine with the water from the still. The propeller now resides in the Central Australian Aviation Museum in Alice Springs, where I have every intention of visiting it.

# 15. Lake Eyre:
## 4,500 Feet Above Sea Level

*The sky. Not so much an empty space, but a soup. A soup of myriad impediments. Water vapour, birds, high-flying insects, dust, gases, flying ice, thermal risings, pollen grains. And on very especial and portentous occasions, a meteor racing itself to extinction. Or frogs surprised by a cyclone. Or a passing soul. Or a falling angel. Or an aviator.*

—Peter Greenaway

**Saturday dawns bright and clear and cool. The birds make sure we're** up early. The galahs—a medium-sized rose-breasted cockatoo, if you prefer—are social, raucous, and comic. They're known for doing things like flying so fast and with such abandon into a tree that they end up clinging to a branch upside down. The corellas are a slightly larger cockatoo, white with a pink cast, and tend to congregate in flocks of hundreds or thousands. Larger, louder, and at first appearing more dignified than their cousins, they are not only contentious over the issue of who gets what branch, but absolutely vocal about it all.

We manage to pack up camp around 9:00 and drive back to Marree to meet Ben Newton at the store. He had hoped to round up some other passengers for our flight over Lake Eyre from the tourists wandering through town, but hasn't had any luck, so we head over to the airstrip. Ben is in his early twenties, and first flew when his dad gave him a one-hour trial lesson as a sixteenth birthday gift. Despite both his grandfather and father being pilots, he hadn't been interested in flying until the first time he took the controls, and then he was hooked. It took him four years to get his license in Perth, working full-time while doing so in order to pay for lessons, but now he quite happily makes a living at it.

The airstrip is more than 4,500 feet long and paved, which surprises me until Ben says that the Flying Doctors use it, as well as his charter service. He already has his six-seater Cessna 206 fueled and warmed up. We do a weigh-in as required by regs, then board, me in front, Kim in the second row where she can roam from side to side to photograph. Ben's only had five flights during the last ten days, and is hoping business picks up in June and July. It's obvious he's glad to fly this morning, not just for the business, but simply to be in the air. We taxi down the runway, turn around north into wind, and take off. He doesn't need much of the runway. It's a powerful little plane.

The first thing that's clear to Kim and me, besides the grid of town, is the old Ghan line and the dirt tracks, the long lines of which stretch over the horizon. Our view is over a relatively flat 360° for hundreds of square miles. I spot water bores here and there, defined by the tracks of cattle coming into them and the same erosion patterns created by the livestock as in both Nevada and the Murray-Darling Basin. Watercourses are carved everywhere, dry for now but marked by banks of darker vegetation. We climb to 4,500 feet, level off, and are making 130 knots into a headwind.

Ben launches into a litany of figures about Lake Eyre from a fact sheet in his lap, some of which agree with what I've read elsewhere, some not. It's difficult to pin down firm statistics on a geographical feature that changes size according to whether it's wet or dry, a place into which all the water from a sixth of a continent drains, yet is the driest part of that continent. The average rainfall around the lake is about five inches per annum, but it can be as low as two or as high as thirty in any given year. And the evaporation rate runs more than eleven feet most years.

Lake Eyre was more than eight feet deep from 130,000 to 90,000 years ago, then completely dry until 60,000 years ago, after which it's been filled from time to time by runoff from monsoonal rains, which fall far to the north. The lake was considered permanently dry until a major flood event in 1949, and a substantial amount of water came into the lake in 1974 and 1984. Although there are playas a little larger in Bolivia and Chile, they never fill completely to become a body of water; Lake Eyre remains the largest ephemeral salt lake in the world. When enough water comes into it, golden perch and several other species of fish migrate down from the rivers and a spawning frenzy ensues, attracting pelicans and other seabirds from Queensland. The perch can grow up to two feet long before the waters become too saline again, and the population crashes, the birds returning north.

Most of Ben's first remarks, however, are about the land speed record set here in 1964 by Donald Campbell, who reached a recorded speed of 403.1 miles per hour in his jet-powered vehicle. That's fast, but the salt crust here is only sixteen to seventeen inches thick at best, and he says that the officials had to channel water into the Madigan Gulf to lay down enough salt for the attempt. Other versions have it that the crust was still wet from recent rains when he raced, which may have slowed him down. In any case, the cars today break the sound barrier and top 760 miles per hour out on Nevada's Black Rock Desert, where the crust is much thicker. The vehicles are so fast that they are by necessity piloted by military fighter jockeys rather than race car drivers.

We cross over the shoreline at the lower end of the Madigan Gulf not far from where Kim and I had parked yesterday, and then fly over the Hunt Peninsula, a closed nature preserve. I note that there seem to be two kinds of dot patterning below us, small ones that are clumps of grass and larger ones that are bushes. Longitudinal dunes run in parallel with each other, but also some larger transverse ones, which I speculate are perhaps relics from an older era.

While still over the peninsula, we spot in the distance where water is flowing down the Warburton Groove, a channel on the lakebed that is the ephemeral extension of Warburton Creek, which enters the lake at its northeastern end. It's a broad channel of water maybe two miles wide. This is the current water front, and we can see where the salt is darker, which marks the larger extent of the flooding several weeks ago. Now we're watching the next surge of runoff work its way down, just one of the pulses the lake may receive each winter.

Ben navigates us out toward the water, farther than he usually flies over the lake—partly because his range is calculated by the charter company to be only an hour long, but also out of caution. If something goes wrong, he doesn't want to be too far from shore. But as we're the only passengers and he's relishing the flying time, he pushes the boundary a bit. I stare out the windscreen, thankful that I'm wearing not just ordinary sunglasses, but glacier glasses, their sidescreens cutting down the glare substantially. To fly over such a large white space in a small plane is to be semi-blinded, cut off from the earth, weightless in mind if not body. Usually when flying this low you can see your progress over the ground, gauge how quickly you are moving over it, but when it's this isotropic you have no guideposts by which to count your way. I try to concentrate on the sights we're slowly approaching.

The dry salt gives way to places where the water has collected and turned red, not from algae, as in the Great Salt Lake, but from the red dirt exposed underneath. Everywhere Kim and I look we see pools of water, some green or blue, some red, some just damp spots of gray or tan discoloration. This is the lowest point in the lake and on the continent, but even given that reference point and with the pools to focus on, I'm still confused. The structure of the lake surface is so complex it might as well be random, and with no idea of the size of anything, there's still no sense of scale. I keep asking Ben how fast we're going, what's our altitude, anything I can use to place myself in this immense and increasingly surreal space.

It's actually a relief when we turn to head back toward the shoreline, heading for the small plateau south of the lakebed where we will take a look at one of the largest artworks on the planet, the Marree Man. Discovered by a charter pilot in 1998, the figure of a naked man raising a throwing stick above his head is 2.6 miles long and was apparently carved by a bulldozer pulling a multi-plow. We parallel the edge of the eroding plateau and Ben spots the figure, then Kim. I keep looking and see nothing until Ben points out a foot as we get further south. Then, as we turn and the light changes, suddenly the entire figure comes into focus. I had been looking at pieces of it all along, but it was way too large for me to grasp. I was looking for something much, much smaller. Its scale is perfect with the lake, visible over its head, but my missing the figure indicates how confused my sense of scale is.

Anonymous faxes sent to a local hotel gave no clue who made the geoglyph, the largest such ground drawing in the world, but did provide clues for the source of the imagery, which is based on a Pitjantjatjara man hunting wallabies. The work has been objected to by some Aborigines, declared an environmental desecration by the government, and when some people came out in 2001 to refurbish the plowed lines, they attempted to erase his testicles and erect penis with a scrawl of lines. Kim mutters, "The Marree Eunuch?"

The figure is fading again not so much by erosion as by plants growing in the lines. There's still no evidence of who made it. Some say U.S. military stationed at nearby Woomera, Australia's version of a nuclear test site, but they wouldn't have had local knowledge of the plateau. Whoever did it would have had to use GPS to track the plowing, and have the plowing equipment. Kim, Mandy, and Guy are all convinced it was local station people. Given the isolation of the site, plus the inevitable gossiping that attends such landscape interventions in the artworld, I'm inclined to agree. One wrinkle is that the Marree Man closely resembles the Cerne Giant in Dorset, England, a geoglyph that was made in the first or second century AD by cutting through vegetation and into the chalk beneath, a common method in prehistoric England. The Cerne figure features an upraised arm brandishing a club, horizontal lines on the upper torso, and a magnificently erect phallus, all features of the Marree Man. For both figures to have such formal similarities by accident seems improbable.

We head back for the remaining twenty or so miles to Marree. When I ask Ben about the number of private aircraft in the country, he replies: "Oh, there's only about three thousand general aviation planes registered in Australia. A single airshow in America will attract as many planes as that in a single day." I ponder the idea that despite the low numbers, Australia is in some ways the most aerialized of all the continents. The Royal Flying Doctor Service, RFDS for short, is a good example of what I mean.

Kim's mother, Marie Mahood, is considered one of the few authentic outback writers in the country, and she was asked by a publisher to write a series of short essays about legendary characters, people other than the famous explorers who had a major hand in shaping the interior. One of them she chose was Reverend John Flynn, the missionary who started the RFDS that's now headquartered in Queensland, but with bases around the country.

Flynn was a young missionary based in South Australia in the years before World War I, and did his rounds at first via camelback from Oodnadatta, a small town northwest of Lake Eyre. His month-long circuits down to Marree and up to Alice Springs gave him firsthand knowledge of how people's lives could be saved if they had been able to reach medical care in time. Solving that problem for the people living in remote Australia became his passion, and the first step was establishing a hospital at Oodnadatta in 1912, then a second one at Alice. He added a used Dodge automobile, then in 1928 convinced Quantas to donate a Victory DH 50. Its maiden flight in the Aerial Medical Service was made in May 1928, the start of the world's first airborne ambulance service. By the end of its first year, "the Victory flew 20,000 miles, the doctor saw 250 patients, saved 10 lives and visited 26 different centres," according to Marie Mahood.

Upon the invention of the transceiver by South Australian Alfred Traeger—a wireless rig small enough to use in the bush that could transmit as well as receive, and was soon used worldwide—the RFDS became the literal lifeline for the interior of the continent. According to the organization's annual report for 2005, it then had fifty planes that flew on average every day more than 33,200 miles and made 159 landings. Its doctors saw 642 patients, made 91 evacuations and hospital transfers, and conducted 202 telehealth sessions by radio. Our stats for the day are more modest. We were in the air for about an hour and a half and logged 158 air miles, but add together all the flights in Australia for mustering, transportation, sightseeing, and the ongoing services of RFDS, and you get an idea of why it's an aerial continent.

After thanking Ben and watering Pirate, Kim and I continue on our way to the opal mining town of Coober Pedy, where we'll stay in an underground motel to balance off my seeing the world from above. As we drive by the edge of the plateau where Marree Man lies, I attempt to take notes while she negotiates the Oodnadatta Track, which in places is as broad and smooth as a two-lane highway, and at others is washboarded, sandy, cut by old floods, patched by bulldozers, and in general a far more interesting route toward Alice Springs than the Stuart Highway, the single paved road the bisects the country from south to north. It's a nice enough road, but most of its traffic consists of "gray nomads," the retired folks with their caravans hitched to a variety of underpowered automobiles. Nice people, the ones we've met, but having to spend all day passing them is a less than peaceful way to see the country.

There was a time when few people would have thought to vacation anywhere in the country but on the coast. The nineteenth century saw numerous attempts to impose the British colonial grid upon Australia by crisscrossing it from south to north, west to east, and although explorations eventually pushed through such routes, it was at a cost. There was no great inland sea to be found, no fertile crescent, no great river, but instead this series of enormous dry lakes, some of the most extensive dune fields in the world, and spinifex, the spiny grass capable of puncturing boots and tires, and that in some places comprised 96 percent of all the biomass within sight. Roslynn Haynes in her cultural history of the "Centre" writes that after the Burke and Wills expedition, artists turned their back on the desert for sixty years, part of a more general failure of Europeans trying to make sense out of it, to assimilate such an alien landscape.

Following the dispiriting failure of the explorers, artists avoided the interior as a source of aesthetic inspiration until 1926, when artist Hans Heysen started going into the Flinders Range. Over the next several years he produced a body of studio work based on field sketches that began to redeem the interior as a source of imagery important to Euro-Australians. Heysen, who was born in 1877 and had been a student of Julian Ashton, the founder of the school Kim Mahood attended, was a superb draftsman and early conservationist. Seen as a rising talent in landscape art, his patrons raised money so he could study in Paris and travel throughout Europe. While there he had adopted the methods of *plein air* painters directly preceding the Impressionists, artists such as Corot.

Heysen was perhaps the last great traditional landscape painter of the country, due in part to his acceptance of the gum tree in all its various glories as an icon of a continent worth painting. Art connoisseurs in Australia knew instinctively that those paintings were destined to be part of the patrimony of Australian art, so when he went into the desert to paint, they were able to accept the subject matter and support his work. He painted the mountains with absolutely no atmospheric perspective, bluing them away into a scale humans are adapted to perceive. It was his way of depicting how aridity diminishes our ability to distinguish foreground from middle- and background. I can attest from our drive the day before that he got it right, but more importantly for the history of Australian art, his almost hyperrealistic style opened the door for the Aboriginal artist Albert Namatjira and the surrealist Russell Drysdale, who were following different visions of the desert landscape.

Namatjira was an Arrernte man who was born in 1902 and died in 1959, his shortened life a distinct contrast with Heysen who lived to be ninety, evidence of more than just genetic differences, but also the gaps between the two cultures in health, nutrition, and lifestyle. Namatjira lived at the Hermannsburg mission to the west of Alice Springs, and in 1934 he watched two visiting painters work. He asked for some watercolors of his own, but didn't have much success until one of the painters, Rex Battarbee, returned and hired him to be a local guide. Their two months in the field gave Namatjira the skills he needed, and he held his first solo exhibition in 1938, which sold out. He became what then passed for an instant celebrity, the next year becoming the first Aboriginal person to sell a work of art to an art museum, versus an ethnographic one.

Based on his acceptance as an artist, albeit in a Western tradition, in 1957 Namatjira and his wife were the first Aborigines to be granted Australian citizenship, which allowed him to buy property in Alice Springs. It's a complicated story, but local resentment landed him in jail shortly thereafter. After serving his six-month sentence he no longer painted and died within months. Russell Drysdale, who was born in 1912 and died in 1981, studied in Europe during the 1930s and although of the same generation of Australia's most internationally known painter, Sidney Nolan, achieved fame a bit earlier. While Namatjira was busy painting some two thousand works, Drysdale over his lifetime painted only a fraction as many canvases. Scrupulous and deliberate, whether done in a heightened realist style akin to Edward Hopper's, or a vividly bizarre one like Salvador Dalí's, his pictures betray that existential loneliness known only to residents of the world's great empty spaces.

Historian Roslynn Haynes has it right when she says that most people's first encounter with the desert here is from an airplane, but it's not just the direct flying experience upon which people base their impressions of the interior. Among the reasons why both the Stuart Highway and the Oodnadatta track are traversed by contemporary Australians has much to do with the aerial imaginations of both Aboriginal and Euro-Australian artists, which are intertwined. The aerial view is found throughout the major currents of landscape imagery created during the mid-twentieth century by the painters who followed these earlier figures. I'm thinking in particular of Sidney Nolan, Fred Williams, and John Olsen, all of whom painted aerial views of desert; and I'm thinking of the Aboriginal dot painters who would turn to their own visual traditions to depict the country, choosing not to adopt the European paradigm.

Haynes states that Nolan is the first known artist to paint the country from the air, but she's not counting the remarkable picture by Margaret Preston done in 1942. Preston was born in 1875 and is widely acknowledged as one of the most important European women painters in the country. In 1925 she started to develop a distinctly Australian modernism, and used Aboriginal motifs to do so, because she thought that the imagery of such a movement should be unique and indigenous to the country. It wasn't useful to Preston to adopt European landscape imagery, and she claimed that only from Aboriginal imagery could a truly national art develop. While the indigenous designs she popularized were appropriated out of context when printed on tea towels and stationary, they prepared Australians to accept an alien visual culture.

Preston painted her oil on canvas *Flying over the Shoalhaven River* in a palette that used colors typically found in Aboriginal art, of which she was also an early and vocal supporter. White clouds, a half dozen or so, float in between us and the ground, while a more solid rank of them waits at the horizon. The sky is only a thin portion of the overall picture plane, which would also become true with Nolan's views. Across her muted ground colors the vegetation appears in what could be taken for an oblique version of Aboriginal dots in their acrylic paintings. In the middle foreground a river meanders upwards and disappears around a bend to the right in the distance. Preston's idea of appropriating the visual vocabulary of indigenous art makes Haynes cringe, even as she admits it helped Euro-Australians see Aboriginal art as more than an ethnographic curiosity. I find that one aerial painting of hers, however, an impressive cognitive as well as a cultural one.

Sidney Nolan was born in 1917 and grew up in Melbourne, where he took night classes at the National Gallery of Victoria's art school in the mid-1930s while working as a commercial artist. He seems to have spent more of his time at the public library than in school, reading Rimbaud and Blake, and looking at the work of European artists, including Klee, Henri Rousseau, Matisse, Picasso, and both the Dadists and the surrealists, influences from all of whom are evident throughout his life's work. In 1938 he decided to become a full-time artist and had his first one-man show two years later in Melbourne. In 1945 he began work on his most famous series, one based on the outlaw Ned Kelly. In this painting, he flattened the pictorial ground, providing a graphic stage upon which his iconic figure would display the obvious influence of his commercial art background and the influence of Rousseau's "primitive" style. As with his French predecessor, the paintings look simple, childlike even, and are images highly encoded with both cultural and personal baggage, not least of which is the story of an outsider in the bush who is brought down by social conventions. What's always been of more interest to me, however, are the backgrounds, Nolan showing an increasing willingness to show the desert as a vast empty ochre stage.

In July 1947 Nolan took his first airplane ride, heading north across the country to Queensland. The flight was an epiphany for him, much of it spent with his nose pressed to the Plexiglas. Afterwards he wrote that it was one of the best ways to understand his country, its structure and colors. During the next six months he took notes while traveling around the northern part of the continent, jottings about a series of panoramic views that he could paint from aerial perspectives both low and high. Nolan returned from Queensland not to Melbourne, but to Sydney where his new paintings—again, narrative ones with figures in them, but with increasingly strong landscape settings—established his reputation there. Kenneth Clark, the legendary art historian, critic, and professor at Oxford, happened across Nolan's painting of an abandoned mine while visiting Sydney, was smitten by what he considered an authentic Australian genius, and became a lifelong champion of the artist. The older man encouraged Nolan to follow his desire to travel inland to paint. Proceeds from sales of the Queensland paintings enabled Nolan, his wife Cynthia, and their daughter Jinx to fly around the interior, which was—and is still today—as expensive as traveling to Europe.

In June 1949 the Nolans flew to Adelaide, and Cynthia made note that the landscapes below them, the squared-off paddocks and contour plowing, reminded her of Klee's paintings. They took a train to Alice Springs, whereupon Nolan hired Eddie Connellan to fly them around. The legendary pilot had pioneered airmail delivery in rural Australia in 1939 and was still using a prewar biplane for some of his mail runs. Nolan climbed aboard with his camera for the four-hundred-mile run from Alice to Murray River, part of a biweekly route that covered eight thousand miles each circuit. Nolan had used photographs for earlier paintings, often historical ones for the Ned Kelly and other series, but this time he would use his own photos. He had modified his medium-format camera with a special mask to create a panoramic format, which he thought would be a frame fitted to the immensity of the desert from ten thousand feet above the ground.

While in Alice he also pondered the bark paintings on display and found Aboriginal art to be a source of inspiration, not so much for any direct technique, but for its relationship to the land, a rare viewpoint at the time. On July 4th the Nolans flew around what was still called Ayers Rock, an experience that he said made him feel "like a bird." When they ended their travels back in Sydney in September, they had flown, driven, and sailed over thousands of miles, and brought back hundreds of photographic negatives and sketches. The job ahead of Nolan was so daunting that he decided to spread out over several years the process of looking at both his own and professional aerial photographs before painting them. From the military he requested copies of their aerial survey photos from above his routes, and lingered over commercial magazine photos of the same terrain. He contemplated how the ground, abstracted from ten thousand feet up and without a horizon when photographed, looked like the Aboriginal art he had seen.

During the war Nolan had switched from rare and expensive oil paints to a synthetic enamel called Ripolen that required him to paint horizontally on the Masonite boards that he favored, often cut to dimensions of roughly 4 x 5 feet. He thus painted standing over his work surface, which further flattened his perspective, an appropriate effect for aerial paintings. By 1950 he had the first thirty or so paintings to show, all but one of them devoid of as much as a single human being. When his exhibition of red, yellow, brown, and ochre aerial landscapes opened, their narrow strip of sky at the top beckoned the viewers into a deep and mysterious territory never before seen by most of them. The renowned anthropologist Charles Pountford introduced them by saying that Nolan had finally invented for Australians an idiom in painting appropriate to the actual land.

Nolan's paintings remain a revelation. Your first thought upon seeing one is that it is a surrealistic landscape, a piece of science fiction that's detached itself from a film and come to rest on a wall. Sere ridges ruck and fold up the surface of the Earth in endless rows, and oddly truncated peaks are eroding, almost melting over time, into the ground. The actual painted surface of the boards is difficult to apprehend at first, so overpowering are the strong reds and so surreal is the land. But when you do approach closely, the image dissolves into nothing but the evidence of hand, motion, pigment, light. It is only when you back away a bit that any of it makes an image. The landscape Nolan was picturing had been ignored not just by artists, but by most Australians except the Aborigines who lived there. As Robert Hughes noted in his introduction to *Recent Australian Painting* (1961), the green valleys of Australia don't extend very far, and as long as the artists avoided the desert as a subject matter, they didn't have much to paint. Nolan had found a way around that.

The next major painter to adopt an aerial vantage point from which to consider the desert was Fred Williams, who was born in 1927, trained as a figurative painter in art school, and went to London from 1952–1956, where he worked at Savage's, a prominent framing shop handling paintings by Renoir, one of his favorite artists. While Williams admired English landscape painters such as Turner and Constable, he was also influenced by continental artists from Claude and Poussin to Cezanne and Picasso. When the young artist returned to Australia, his immediate reaction upon setting foot ashore was that he wanted to paint the landscape and gum trees. They were already aesthetic clichés for the country, given the works by Heysen and Nolan, but he felt he could do something new with the landscape both as a subject and as a ground for formal aesthetic advances.

Williams tended to work from layers of paint thinned with turpentine to ones increasingly loaded with oil, which meant a canvas could take six to twelve months to finish. But that also meant he could work simultaneously on several paintings in several series, so he had a chance to build upon his experiments from one canvas to another. In the late 1950s he abandoned the conventional landscape canvas proportions of one vertical to one-and-a-half horizontal for square canvases, and in 1959 started tilting the ground of the painting towards the viewer, applying the Cubist still-life technique of Braque and Picasso to landscape. In the early 1960s he settled on a 5 x 7 foot size for his canvas, and he was dividing the "ground" of subject and object with clumps of rich impasto marks in clumps, the vegetation of windbreaks dividing the plane into golden sections. Williams tried at times to include people and buildings in his compositions, but he wasn't pleased with the results. In part that may have been because the perspective he employed was like that in the Chinese paintings he so admired: what was far away appeared at the top of the canvas. That's a scheme that works better for landscape than peopled scenes.

In 1967 Williams visited the arid interior of Australia for the first time, and was stunned by seeing the bony structure of the land laid bare. He gained a sense that the continent had a unified geological reality underneath its various landscapes. In response, he decided to drop place names and regional references in order that his "Australian Landscape Series" (1969) would be taken to represent the whole of the country. The severe reductionism of the series, which Williams acknowledged was indebted to both maps and Aboriginal art, was an effort to convey the size of land, its vast scope. And he deployed an aerial viewpoint that emphasized the fractal nature of the land. In a Williams painting you often don't know if you're looking at something from very close up or from very far away, which is how land is structured by natural processes, self-similar at all scales. To have arrived at such a conclusion intuitively and through a formal consideration of aesthetics is one of the great imaginative leaps of landscape art in the twentieth century. Patrick McCaughey, the Melbourne art critic, has noted that because the Australian landscape is in places relentlessly repetitive, it led Williams naturally to set down small marks on large canvases. Patterned markmaking, which in other contexts was seen as abstract gesture, was here revealed to be representative.

By the late 1960s Williams was continually sketching outdoors, taking trips to Broken Hill and elsewhere with other artists, such as John Olsen. Of relevance to his aerial perspective were sojourns in 1975 to Mount Kozzie, including a hike to its summit from Charlotte Pass, and the Werribee Gorge. His painting *Mt. Kosciusko*, a large oil completed the following year, relied upon photographs he had taken from airplane flights between Melbourne and Canberra. When painting the canvases in his *Guthega* series, pictures of snow in a gully seen from a track above it, he worked on his canvases both rightside up and upside down, as if reading the landscape coming and going along the track. The Werribee series, which started with a large panoramic view four feet high by fifteen feet long, evolved into an aerial series with no horizon line, and showed the Snowy River far below, as much a powerful meander of paint as a landscape feature.

Finally, in the painting *Snowy River at Guthega*, painted while he was working on both series, Williams achieved a flat abstract field of view that resembles an Aboriginal dot painting as much as it does a landscape done by a painter influenced by the French Impressionists, and McCaughey noted its haptic qualities. Williams said he was exploring the "flesh" of the landscape in these paintings, and it's not coincidental that Aboriginal acrylic landscape paintings stem from finger painting on bodies and sand drawings or paintings done by pushing pigment into the ground. Williams was increasingly attracted to the view from atop hills and aerial photographs in newspaper. In 1975 he clipped an aerial photo of cattle stranded in a line atop a levee bordering the flooded Murray River, and from it painted a canvas roughly 101 inches long by 24 inches tall. He worked on this eccentric painting both horizontally and vertically, and this willingness to work on aerial view from a variety of directions resonates strongly with the work of Aborigines who both draw in sand and paint with acrylics from multiple directions.

In 1976 he saw a brushfire from a commercial airliner while flying to China and Europe, and immediately upon arriving in Paris painted watercolor sketches of the fire front along a sixty-mile-long line. The twelve gouaches, *Brushfire in Northern Territory*, were his first aerial series based on personal views from an aircraft. The next year Williams became the first Australian artist to have a solo exhibition at the Museum of Modern Art in New York City. Upon his return to Australia, he received an invitation from the sponsor of the exhibition, the mining company Comalco Limited, to visit their operations at Weipa on the Cape York Peninsula in Queensland. His flight in the company plane gave him what curator and art historian James Mollison called Williams's "first really extensive view of the land from a light plane." He was stunned and "used a mapmaker's view of the Cape York Peninsula to describe the infinite space of the outback." A river below is painted "as if it were an Aboriginal mark on the landscape."

Williams showed the Weipa paintings to the chairman of Conzinc Rio Tinto, who promptly offered to get him in an airplane over their iron mines in Pilbara in Western Australia in order for him to paint a series the corporation could purchase. The flights took place in May 1979, and Williams asked the pilots to fly him repeatedly over areas that interested him so he could take numerous photographs from which to work. He completed almost a hundred gouaches based on his first five-day visit in May and a subsequent one over eight days in June, paintings that examined the Pilbara from intimate studies of spider webs and flowers up through aerial sketches. It wasn't until March through June of 1981, however, that he was able to create the thirty-one larger oil paintings in the *Pilbara Series*. By then he was painting quickly and with little revision, using rags as well as brushes to get down his vision.

Williams painted his last works in 1981, including *Winjana Gorge, Kimberley, I,* that hangs in the Metropolitan Museum of Art in New York

(Plate 9) before dying the next year at age fifty-four of inoperable lung cancer. He was still working from his Pilbara photos, aerials of tidal flats and salt pans. Williams had long desired to create a uniquely Australian landscape vocabulary, and at the end of his life had decided that the aerial view provided exactly that, a vocabulary that had ties to the marks and perspective found in the Aboriginal art that was then gaining increasing prominence. In 2001 the mining company Rio Tinto donated the original *Pilbara Series* to the National Gallery of Victoria, and when I visited Melbourne eight of the canvases were on view, including the great *Iron Ore Landscape* from his last working year, a picture in which the horizon of the red landscape with its dots of brush is tilted steeply down to the left, a view of the landscape as the pilot banked sharply into a turn.

The New Zealand writer and painter Gregory O'Brien, who has written about the aerial imagination in Australian painting, suggests that Williams's work "echoed . . . the pictorial strategies of Post-Painterly Abstraction as embodied by the work of Barnett Newman, Mark Rothko and Ad Reinhardt." In walking through the Melbourne collection and examining his paintings with former curator Grazia Gunn, it seemed clear to me that she ranked Williams's work equally within Australian landscapes and twentieth-century abstraction, and I suspect that the reason Williams was favored with a show at the Museum of Modern Art in New York was that his work made sense to curators accustomed to the abstractions of the painters mentioned by O'Brien.

John Olsen was born in New South Wales a year after Williams, attended the Ashton Art School in 1950, and by 1955 was having his first major show. Olsen, who knew Sidney Nolan, started painting outside with Williams and others in 1969 when living in Dunmoochin just outside Melbourne. Sometimes he would lay his canvas on the ground and paint from above, which he had observed Williams doing. The environment there was dusty hills and eucalypt, stock ponds behind circular dams, and the paintings are devoid of figures, a high horizon line flattening out the landscape. The influence of Williams is obvious, and like the older artist, Olsen was steeping himself in Asian art and literature. He was reading Suzuki on Zen, thinking about calligraphic painting and Sengai brush drawings, pondering the poems of Basho. In 1981 Olsen published a book

with journalist Sandra McGrath titled *The Artist & the Desert*. In the introduction Olsen evokes the saying of a classical Chinese scholar about the
best way to view a landscape, given its scope, was from the mountains,
for only from there could the details be seen.

In the early 1970s Olsen began to visit Lake Eyre, becoming the first
artist to use it as a subject matter in other than a topographical manner. In
1974 the environmentalist Vincent Serventy told him about the rare large
flood on the lake with its attendant boom in wildlife, and convinced Olsen
to accompany him on what would be the first of his three trips to the
playa with the naturalist. In 1984 Olsen and Serventy, along with historian
Mary Durack, the poet Geoffrey Dutton, and photographer Alex Bortignon
led an expedition to Eyre, a model for the trips into the desert organized
now by Mandy Martin and others. One of the Olsen paintings in the book
*The Land Beyond Time*, which documents the expedition, pays homage to
the role of the aerial. *Spinifex and Aeroplane*, a diptych, is painted from a
vantage point high enough to show a small airplane flying below the
viewer yet over the landscape.

In 2001 Olsen was part of a group of ten artists camped out at William
Creek. They took a helicopter to establish a remote camp an hour to the
north and west of the Warburton Groove. Although photographer Hari
Ho spent much of his aerial time taking pictures of the lake from 5,500 feet,
he also took a photo of David Dunn, their pilot, standing in front of an airplane propeller from a crash caused by a man flying below sea level when
his instruments failed, a picture that always makes me wonder about that
story of carving a propeller out on the playa.

Other artists were inspired by Williams and Olsen to use aerial views
of the desert, including Robert Juniper who was particularly fascinated
with dry lakes in his native Western Australia, and the Sydney artist John
Colburn flying over the Nullabor Plain. Juniper's work is a clear inheritor
of the style pioneered by Olsen; Colburn's aerial work tends toward
totemic geometric figures, a private symbolism placed on a monochromatic field that is so abstracted away from the desert forms that we can no
longer discern its connection to topography. Perhaps the most widely reproduced Australian aerial painting is by Tim Storrier, an artist two decades
younger than Olsen who accompanied the older artist and Serventy to Lake

Eyre in 1976. His high-altitude view of the desert from a jetliner, *An Alti-tude* (1977) remains a signature image for the country. The six-and-a-half-foot square canvas is painted predominately in tans and browns, the horizon a thin band at the top, wispy clouds scattered at random over the great parallel dunes of the interior. A scrim of dots overlays the complex watercourses braiding in between the dunes. Like the paintings of Williams, the resemblance to an Aboriginal dot painting is inescapable.

That the most prominent Australian painters of the twentieth century chose to paint their country's landscape, and to do so from the air, tells us a great deal. First, despite the fact that Australia is the most urbanized continent, by far the majority of its population living along its coastline, the interior remains a source of metaphorical fascination, as well as pro-viding water, mineral wealth, and agricultural product. Second, the aerial vantage point is a useful one from which to observe such a large and rel-atively flat space. Third, there is a connection to tradition based on the early topographical conventions of Europeans surveying the land. None of that, however, explains why it's aerial painting instead of photography that displays this fascination.

Yes, there are Australian photographers working from the air, most notably Richard Woldendorp, who, along with his own views, has delib-erately photographed from the air some of the subjects painted from that perspective by Williams and Olsen. His 1981 photo of a "scrub fire" from the air could almost be mistaken for one of Williams's 1976 paintings of the same phenomenon. But I think the connection for Australians is much deeper than that, one rooted in the visual culture of the Aboriginal people and the ways in which their sand paintings, and since the 1970s their dot paintings, depict the land.

Kim and are talking about all of this as we leave the Eyre Basin and head west toward Coober Pedy, at one point cutting across a corner of the Woomera Prohibited Area, at 49,000 square miles the largest test range in the world (by comparison, the Nevada Test Site is only a paltry 1,350 square miles). I ask her about whether or not her Aboriginal artist friends fly much. She's amused at my naiveté.

"Look, some of the more well-known artists charter airplanes and fly around. It's kind of an ironic reversal, the white artists trying to make do on the smell of an oily rag," Kim says with a grin. "But the aerial is nothing new to them. Their ancestors were coming up out of the ground and flying all over the place. They dream about it and there's not much separation for them between dreaming and reality. So Aborigines were flying about by whatever means you want to call it before there were airplanes."

# 16. Alice Springs

*When I looked at the earth, I saw all the lines of cubism made at a time when not any painter had ever gone up in an airplane. I saw there on the earth the mingling lines of Picasso, coming and going, developing and destroying themselves, I saw the simple solutions of Braque, yes and once more I knew that a creator is contemporary, he understands what is contemporary when the contemporaries do not yet know it, but he is contemporary and as the twentieth century is a century which sees the earth as no one has ever seen it, the earth has a splendour that it never has had.*
*—Gertrude Stein*

**The footbridge along Undoolya Road is a long span crossing the Todd** River, a watercourse that at this time of year is bone dry. The elevated pathway gives me a chance to meet people as I walk from my motel into downtown, to check out the astonishing variety of vehicles on the road below, and to admire the white gums growing in the sandy bottom of the channel, as well as the large river red gums along the banks. The latter are nicknamed "widow makers" for their habit of dropping large branches unexpectedly, and Kim and I have avoided camping under them for that reason. A few branches rest in the broad streambed, part of a natural system for slowing the floodwaters that pump through here every few years.

Driving into Alice Springs with Kim a few days ago had been a revelation after almost two thousand miles in the truck, many of them trailing clouds of dust behind us on dirt roads. Once you cross the border between South Australia and Northern Australia on Highway 87, the vegetation is lusher, sandstone outcrops define the contours of the land, and the ground turns from dust to sand. The red granules contain high proportions of silica and hematite, and have weathered over the millennia down to spheres about one hundred microns in diameter, all of which makes the sand sparkle against the blue sky, the hues of each heightening the other. Even the salt pans are red.

The highway into Alice cuts through the MacDonnell Ranges in the Heavitree Gap, a narrow slot in the abrupt sandstone ridge. The road, telephone lines, the Todd River, even the railroad tracks squeeze through the singular gap. The town was startlingly green after the desert. "There were never so many trees here when I was a kid," observes Kim. Bougainvillea and oleander are blooming in many of the yards we pass. When Kim was growing up here in the 1960s, the town had three thousand residents; now it's more than twenty-five thousand. We pass through downtown, then cut right to get to my motel and a few blocks away the house where she'll be staying, home of the artist Pam Lofts. The park across from her primary school was dirt, but now it's a field of grass. A fifteen-year drought ended here in the 1970s, about the time the Mahood family left "the Centre," and the trees have gotten huge in the meantime. "There used to be dust storms here that would boil up out of desert, hit the mountains on the other side, then arc over town before the sand collapsed into the streets. That wouldn't happen now; there's too much vegetation out there. The mulgas we've been seeing are all about thirty-five years old, and date from the end of that drought."

As I've been walking back and forth through the residential neighborhood from my motel to Pam's, I pass solid metal fences about six feet tall wrapped around entire lots. I asked her when people started putting them up, and Pam laughed nervously. "1967," she said, implying it was a reaction to the referendum that gave Aborigines citizenship. She was half serious. Everyone uses them for their backyards; as they're a cheap and practical windscreen. But walling off the front of your house to the street was and is often a response to having Aboriginal people invade your space. Pam's yard is open to the street, but she told me: "I've woken up and found people sleeping in my verandah, and for a single woman, that's unnerving."

Pam's a middle-aged sculptor who's just returned to Alice this March from three years in Canberra getting her master's degree in art at the Australian National University. Kim and I went with her one afternoon to move a work she had done in the late 1990s, a line of poetry carved into individual slabs of sandstone with different typefaces. It's been sitting in the Desert Park, a rare work of Euro-Australian art in a nature preserve encompassing five square miles of carefully tended native plants and animals. It's time for her to bring it home, but first she wanted to photograph it on a small knoll with the MacDonnells in the background.

"tREAD *softly* because YOU tread on my[YOUR/our] dreams," it reads, a play on entwined notions of environmentalism and sovereignty that manages to evoke Aboriginal rights, the colonial past, and the current political climate of the country.

I've been spending my days visiting the local museums and galleries, meeting with artists such as Pam Lofts and Mike Gillam, a photographer who's done aerial work over this part of Australia. Mike and his wife, Maria Giacon, live on a property once owned by a welder and junk sculptor. You enter their place by passing through an iron gate wide enough to accommodate a bus, an entrance mounted between a pair of tamarisk trees with trunks larger around than most automobiles. Tamarisk, or salt cedar as it's sometimes called in the United States, is a dense evergreen native to neither country, and was imported to slow erosion along watercourses. It's salt tolerant, has a long taproot that can drain groundwater faster than you can empty your bathtub at home, and grows in great choking thickets wherever it can. Most people in Australia would like to see the species eradicated from the country, but these two trees are so beloved by the community that people won't let Mike cut them down. They also provide habitat here, as in the American Southwest, to a variety of bird life.

Mike's continued the tradition of assembling found objects into art, the largest example of which is the legendary Silver Bullet Cafe, which is sadly now open only for special occasions. The restaurant is made from several old and very handsome aluminum Silver Bullet trailers—think

Airstreams—that were once used for remote area classrooms. Mike toured me around the property, all the original pieces still in place, along with more recent additions by artists he's invited to work there. It looked as if he had recycled much of the heavy machinery in town. Birds flitted around happily, and at one point he got down on his hands and knees to pull up some duff so a small and friendly wagtail could rummage around in it, its trademark plumage standing straight up and semaphoring happiness from side to side.

Mike worked for the Aussie Parks & Wildlife for seven-and-a-half years until 1981, first as a research technician, then as a ranger dealing with weeds, feral cats, and law enforcement issues such as poaching. His work reflects that background, whether it's close-ups of the local caterpillars that were long a staple of the Aboriginal diet, or wide aerials shot from a helicopter using a gyrostabilizer with his medium-format cameras.

"The big red one" is what he called the panoramic aerial view more formally titled *Shifting Ground* taken south of town. "Someone who knows how to read the ground will know what's meant. That unnecessary track someone has made that's already turning into an erosion gully? That will become a watercourse, buffel grass will move in. Roads are destruction. I took it in 2003 during a drought year. There's ash on the ground from a fire, you can see how it fell. All the regrowth you see is just from moisture in the ground. It's an utterly suspended moment. I know that if you thumped the trees, the leaves would fall." Almost as an afterthought he added, "Alice has the highest rate of per capita water consumption in the country, but we're not doing anything about it."

Mike plans his flights carefully, in part because he can't afford more than about one a year, but also because he worries about the carbon footprint necessary for obtaining just one good photo. The *Desert Oaks at Sunset* is a photo taken from a fixed wing about four to five hundred feet up in the air. It's one of those aerials that confuses you in some subtle way and he pointed out why. "See the shadows, they go upwards in the photo, away from the viewer, which we normally don't see unless we're elevated. The only way to see a shadow on the ground is if it's coming toward you—so it confuses people." And I realized, as he said it, this was one of the reasons I'd been so disoriented by Mike Light's photo from the

White Mountains above the Mojave desert. The shadows were pointing the wrong way.

People say Gillam's work is all about pattern, but I think it's more about relationships in nature. The patterns are just a manifestation of those relationships. He pointed again to the *Desert Oaks* photo, commenting that their canopies "are like little green clouds floating over the ground, and underneath them there's a leaf mat, which changes the chemistry of the soil, so the rings of spinifex don't get too close." He paused, looked at me, shook his head. "Do you know how dishonest a picture this is? How I had to find a place without tracks and fences and bores and cattle?"

So while I'm standing here on the footbridge, I'm thinking about the aerial landscape, pattern and nature, and the acrylic dot paintings hanging in the Papunya Tula gallery downtown. And I'm watching Aboriginal kids playing in the riverbed, wondering how many of them know about the rock by the causeway north of the footbridge called *Atnelkentyarliweke Athirnte*. According to *Site Seeing*, the book about sacred sites in the Alice area, the rock is a Dreaming manifestation of the caterpillars who camped here overnight, caterpillars being the most prominent ancestral being for the area. The caterpillars painted themselves and established the ceremonies and laws by which the people, who were not yet on Earth, would ensure the continuance of their species, a mutually beneficial arrangement, given the high protein value of the insects. "The older Arrernte people of Alice Springs remember how, in the early days of the town, this rock was always kept covered by river sand, so that it would not be seen by women and uninitiated men," says the book.

Trying to describe Dreamings and the paths they took—the travels of the ancestral beings that created the world—is what Bruce Chatwin wrote about in *The Songlines,* a book that popularized the subject. It's an endeavor I approach with trepidation. People with long strings of capital letters after their names signifying major academic accomplishments have spent decades working with Aboriginal people to understand what has been called the world's most complex social system. I list some of those studies in my bibliography, a list that grows by a title or two or three every week that I'm here, and I would suggest anyone interested in the subject, or baffled by my less than adequate explanations, seek out one or two of them.

Here's how the anthropologist Peter Sutton puts it, after spending several paragraphs telling us what they aren't. There's no distinction between the spiritual and the material with the Dreamings. The natural and the supernatural are one. Furthermore, Dreamings and their manifestations, from waterholes to rock formations to animals, are one. Further yet, if you're not respectful of a painting representing one, it—the painting, the Dreaming—may cause you harm. When we look at a dot painting and see circles and lines, we're looking at a geometricized, usually symmetrical representation of sites and paths between them that represent actual features in the landscape—albeit arranged according to the needs of the design, and not necessarily correlating exactly to the topography.

Do you know that painting of a pipe by Rene Magritte that has the line "Ceci n'est pas une pipe" painted underneath the pipe? "This is not a pipe" is literally and quite famously true, as it is merely the picture of a pipe. Well, imagine if Magritte instead insisted that his painting was not an image of a pipe, but a manifestation of The Pipe itself, a painting that was an act of devotion to an object that was made by the gods, evidence of their passage, and that to paint the pipe was to ensure that tobacco would always grow and produce peace when smoked with other adults at your level of initiation in the tribe. That's a crude and almost completely inaccurate analogy, but it's one that often occurs to me when I'm looking at Aboriginal art.

We see the designs of the paintings and assume they are a view of the ground from above, and they are, viewers are looking down at representations of a Dreaming track or songline on the ground. But they can also be the view from below, as that's where the Dreamings live. The ancestors came out of the ground, created various features, then went back down into it. The paintings represent both country and kinship, thus at the same time a single Dreaming can outline territorial boundaries, land claims, and who can marry whom and from what tribes. The list seems endless because it is. Aboriginal art, as the eminent Australian anthropologist Howard Morphy told me, is a system of knowledge. And it's had considerably longer than ours to evolve. Peter Sutton says that "Classical Aboriginal culture concentrates on human relationships, or on how things lie in relation to one another, rather than on things themselves. Relationships between people (kinship, social organization, amity, conflict), and between people and place (homelands, sacred sites, mythology, songlines, topo-graphic icons)."

It is also true that the designs act as powerful mnemonic devices, rein-forcing the ability of people to remember how to act, where to find resources, when to observe certain rituals. I sometimes think that every landscape feature on earth that's been seen by people has a story to it, a meaning assigned whether it's a personal memory or a collective one. Mike Smith, the anthropologist at the Australian National Museum in Canberra, is another one of the scientists responsible for pushing back the timeline for the Aboriginal occupation of the continent, and we're been discussing how the art of the Aborigines stems from the fact that they're the oldest living remnant, and therefore the most complicated version of a paleolithic hunter-gatherer culture on the planet. He's careful to qualify: "We're not looking at a paleolithic culture, but an evolution from it." Looking at the rocks and thinking about the difficulty of my writing any-thing sensible about Aboriginal culture reminds me of how problematic understanding even much younger native cultures in my own desert Southwest can be.

Keith Basso is a revered anthropologist, a former rancher who has done fieldwork in Australia as well as working extensively with the Western Apaches of Arizona for over three decades. In 1979 he began to study the Apache use of place names, a five-year project out of which he published *Wisdom Sits in Places*, a book about landscape and language that demon-strates the complexity and variety of the representation of place in their culture, and by extension, the complexity of the Aboriginal culture. In one instance Basso recounts a conversation among four Apache women about the misfortune of one of the speaker's brothers, a younger man who had gotten sick a few months after inadvertently stepping on a snakeskin shed among some rocks. An older man reminded him how dangerous snake spirits are, and counseled him to seek the help of a "snake medicine per-son," but the brother had brushed off the advice.

The response of the three friends consisted of nothing but place names: "It happened at Line of White Rocks Extends Up and Out," says one per-son. "Yes, it happened at Whiteness Spreads Out Descending to Water," says another. "Truly. It happened at Trail Extends Across a Red Ridge

With Alder Trees, at this very place!" adds a third. The speakers are evok-
ing stories based on shared remembrances of historical and mythical titles
of stories about events at specific places wherein young people acted fool-
ishly, didn't heed the advice of their elders, and got into trouble. Two of
three stories were chosen deliberately for their humorous tone so that the
sister was not offended, and thus could absorb the gentle advice being
offered. "Talking in names" is what the custom is called, the concatenation of place and memory codified into stories so that the knowledge
gained is preserved, whether it is a moral tale or one regarding how to
find the resources necessary to survive in the desert.

Basso eventually came to realize that for the Apaches wisdom was
based on such careful observation and that in memorization of emplaced
stories a "steadiness of mind" was achieved. With that level of calmness,
a wise man is able to react swiftly and surely to various challenges. If one
is able to maintain a resilient mind as well, and thus reach a stage referred
to as a "stillness" or "smoothness" of mind, one becomes able to foresee
dangerous or troublesome situations. As in most pre-technological cultures, some of the stories linked to place arose from what we would consider the mythical past, a path of history that the Apaches consider a trail
blazed first by their founding ancestors and still walked by living
Apaches today. Some of the places were named, which is to say storied,
by the ancestors, but some were named by local people as recently as a
few months earlier. It is this deep and ongoing construction of place that
allows the Apaches, in their view, to "live right."

Although all desert peoples around the world do not respond to their
arid environments with the same kind of cultural practices, I bear in mind
that all our brains are similar physical structures that, in conjunction with
our likewise similar sensory physiology, tend to produce cognitive
responses that are enough alike around the world that we can classify
them all as human. If Aboriginal people responded to resource scarcities
caused by drought and other factors, and survived by dispersing and
moving over large tracts of terrain, that also meant traversing the territory
of other tribal groups. Having a series of emplaced stories would have
been an effective survival strategy through which to remember the physical and seasonal location of waterholes, edible plants, and migrating

game, as well as a way of negotiating safe passage. Dan Witter, formerly an archeologist with the park system in New South Wales, notes that Dreaming tracks are the largest and most evolved network of routes known in the world, an encoded navigational system-reticulum passed from generation to generation through rituals made while actually walking the tracks (and singing the accompanying stories, hence "songlines.") Dreaming is a system that answers the question of how to live right in a specific environment, Australia's much older equivalent to the Apache "talking in names."

The songs, thousands of them in different languages and dialects reflecting the diversity of the Aboriginal people, lay out an affinial system for marrying across tribes (perhaps meant to preserve genetic diversity), moieties (hierarchical social groups conferring privileges), and reciprocity rules for the exchange of goods and information among groups. The visual culture of the Aborigines reflects the oral culture: physical evidence for the antiquity of the former including ochre used as pigments in thirty-thousand-year-old burials, and painted rock art on Cape York that's twenty thousand years old. The most widespread rock art style on the continent is Panaramitee, which dates from the Pleistocene. That style alone has 130,000 separate motifs, according to anthropologist John Mulvaney. Witter accounts for the complexity of the Dreamings and their manifestations by saying that the Aborigines responded to the challenges of continental aridity and climactic variability with a social innovation instead of a technological one, an idea I find appealing.

The first thing to understand about Aboriginal paintings from the Western Desert is that they are not maps in the sense that we define such objects. Although they may have great metaphorical and narrative consistency, and be faithful to Dreamings, they don't represent terrain in a manner that is consistent internally or externally, nor are they accurate about the spatial relationships of one place to another. Why should they be, since it is not their purpose to serve as a graphic guide to terrain? All humans, it has been proposed by a variety of cognitive scientists and cartographers, may have a hardwired mapping ability—akin to but not the same as what Blaut found in children as the ability to understand aerial photos. Again, aerial photos aren't maps; they are pictures of the ground below you, not representations of it using a symbol system that has to be learned and decoded. Aborigines in Australia display the same abilities in constructing and reading maps as do other people. When explorers during the early years asked them to draw maps, they either did so in the sand, or, once they learned how to use drawing instruments, did so on paper. The results were what you would expect of someone drawing a map for the first time, and they look nothing at all like their paintings. Aboriginal artists have also demonstrated that they are able to make paintings that are realistic landscapes from both ground level and aerial perspectives. Again, that's not the point.

Most of the items made by Aborigines that we'd be tempted to call art have been ephemeral, ritualistic designs put on wooden shields or carved into "scar trees," such as the ones that Mandy Martin and Trish Carroll and I visited near Pennyroyal. Or they were performative by nature, sand drawings done as part of ritualistic storytelling sometimes involving dance and gesture, the designs erased after each episode in the narrative.

The oldest surviving examples of visual culture made by Aborigines are those carved into rock, what many people call rock art, a term that not all anthropologists around the world will readily accept (although the Australians do, for the most part). While I was staying at Pennyroyal, Trish, Mandy, and I visited the Bigga Rock Shelter site not far from the Lachlan River, driving from an art opening of contemporary Aboriginal art at the Cowra Art Gallery in the morning to the top of a hill strewn with huge granite boulders in mid-afternoon. On the dirt road we passed a shingleback goanna trudging sluggishly along, a dark lizard the size of a loaf of bread covered in heavy scales that looked like something from the Jurassic. The site itself, a rare overhang with pictographs tucked underneath, was fenced to keep out vandals. Almost three hundred figures and designs decorated the rock; clay, ochre, and charcoal had been used to create the white, red, and black figures. The density and varying condition of the designs, some faded almost completely away or painted over, suggested that the site had been used for thousands of years.

The major motif of the site and its most prominent design was a long serpentine shape. Trish thought it represented the river, but the interpretive plaque to one side said it could also be the Rainbow Serpent, an important Dreaming figure. Mandy thought it might be standing in for both the river and the serpent. There were anthromorphs, kangaroos, other animals I couldn't identify, possibly a goanna or two. None of the shapes had anything to do with a map, as such, but more with the iconic presentation of site-specific Dreamings.

So I've been wandering about Alice Springs looking at the history of Aboriginal desert painting and thinking about four works in particular, starting with two in the art gallery of the Araluen Cultural Precinct, where it so happens some of Mike Gillam's aerial photographs and paintings by Mandy Martin are hanging, a nice happenstance that is more than coincidence. Among the facilities at Araluen are the art gallery, an anthropology research center and natural history museum, and the Central Australian Aviation Museum. That kind of shows you where their priorities are. The Royal Flying Doctor Service plane *John Flynn*, which flew from 1952 to 1970, sits here across the tarmac from the E. J. Connellen hangar at the site of the original Alice airport that opened in 1921. The first commercial flights were brought to Alice in 1935, and by 1938 Connellen was running a forty-thousand-mile aerial survey of the Northern Territory. Putting art, both Aboriginal and Euro-Australian, next to aviation strikes me as an accurate and revealing juxtaposition.

In the art gallery there's a room devoted to the work of Aboriginal artist Albert Namatjira, whose watercolors started the movement called the Hermannsburg School, after the mission where he was raised. At first consideration the work of Namatjira and others in the movement appears to be straightforward imitations of colonial landscape paintings, perhaps a little more vivid and graphic than Euro-Australians were used to seeing, yet very much within a European tradition. But, as Kim Mahood points out, he was painting the country, the Dreaming stories, just with a different technique. Alison French, the preeminent expert on his work, says that Namatjira was painting not just the sights, but the sacred sites of his people, and using traditional marks related to those found in rock art to do so.

As Kim and I walked through the paintings, she pointed out how the Hermannsburg work in general got more and more hard edged and graphic and abstract over time, and the paintings by Otto Pareroultja are my favorite examples. His *Rockhole Creek*, a watercolor done in 1945 assumes a vantage point high in the MacDonnell Ranges as it follows the course of three waterholes flowing into one another, a dizzying perspective downward as they run along a rocky cliff edge. In the background the vegetation is painted in ridge-like dots, almost like a dot painting in a three-perspective view, versus a plan on the floor. In a later painting, *Waterhole with Mountains*, three successively distant ranges are done in lines and dots, until the last one is nothing but a ranked series of brush dabs. In Pareroultja's late paintings, done in the 1960s until his death in 1973, the landscape is composed solely of small geometrical figures patterned together into totemic landforms.

In the room next door is an exhibition of Papunya art, its ochre walls hung with fifteen classic acrylic dot paintings from 1972 and shortly thereafter. Papunya was established in 1959 by the government as an Aboriginal resettlement community, part of the national assimilation program begun in 1951. Meant to dispossess Aborigines of their heritage in order to acculturate them more thoroughly to whitefella ways, people from a variety of language groups and locations were thrown together in a depressing situation likened more to a concentration camp than a town. The arrival of a young art teacher in 1971 would lead to an artistic movement powerful enough, ironically, to overthrow much of the government's preconceptions and regulations regarding the indigenous population.

The story of how Geoffrey Bardon and the people of Papunya re-empowered Aborigines through art that eventually would be used to argue for their rights is one of the most poignant stories in Australia, in art, and in the history of civil rights. No matter that the paintings have gone from selling for $35 to $2.4 million at auction, the level of commodification never belies the fact that the artworks have played a unique and critical role in helping rescue an entire ensemble of related cultures.

There's a film playing in the gallery, original footage shot by Bardon in Papunya before he was himself exiled from the community by politics. It's lovely to watch the men doing a sand painting on the ground, then the same men creating acrylic paintings—also while sitting on the ground, the materials laid out in front of them. Sand or ground painting is done with the fingers and is a haptic activity: You push at the ground and it pushes back, thus establishing a relationship. One of the hallmarks of desert sand drawings is that the longer the Dreaming entity sat in a place, the deeper the mark was. Time was shown as having a physical depth that was also a tactile dimension.

One of the paintings that's in the film is mounted on the wall to the left of the video monitor, and I can see it out of the corner of my left eye while watching the film, a lovely bit of curation. It's *Children's Bush Tucker Dreaming* by Johnny Warrangkula Tjupurrula, an acrylic-on-board from 1972. Bardon credits Tjupurrula with being one of the first men to "transcribe" images from sand to board, and then to reimagine the continent through constant innovation in his paintings. The Pintupi men and others at Papunya knew painters working in the Namatjira style, but unlike the commercial artists such as Rex Battarbee who taught Namatjira, Bardon encouraged his friends to use their traditional motifs to depict their lands and Dreamings. Johnny Warrangkula Tjupurrula's painting is an important one as it shows the still-new use of dots. The traditional application of dots on the body and in ground paintings was to outline figures, but Warrangkula deployed them as a field in which the narrative occurs, as well as to represent dense vegetation on the ground after a particularly rainy season (the source of "tucker" or food).

Bardon believed that the use of dots slowed down the painter and allowed him to concentrate better on the craft of the work, a desirable qualitative attribute that would fetch more money for the work. Warrangkula used the dots to create sacred ground, not simply a representation of it, the painting being the Dreaming of plenitude, of rich resources. The dots change color in different areas to show the change of the land over seasons, to make apparent all at once a sweep of time. As Kim shows me, they also served to hide the objects of sacred knowledge. In the upper left-hand corner is the head of a "Ceremonial Man," the rest of his body obscured by the dots, which he overpainted after a discussion with his peers led him to realize he was giving away too much sacred knowledge. "An artist wouldn't do that now," says Kim, "it's too revealing."

The use of dots has evolved over time, and there is a synergistic relationship between what white people think they represent, the assumption that an aerial view is implied, and even the density of dots to the value of the paintings, the logical outcome of Bardon's praising their use. This has been true even for Aboriginal artists from other parts of the country and working out of a different material tradition, such as Emily Kame Kngwarray and Kathleen Petyarre who grew up in the Sandover region some 150 miles north-northeast of Alice Springs. The men at Papunya derived their designs primarily from male rituals, whereas the paintings by women arrive through women's knowledge and fabric design. The most famous Sandover artist is Kngwarray, who was born around 1916, but didn't start painting until she was an older woman in 1988. She had only eight years in which to paint and produced perhaps as many as three thousand works in that time, sometimes at the rate of one a day. In 1997 she posthumously was one of several artists representing Australia at the 47th Venice Biennale.

The women of Sandover began experimenting with batik dying of fabric in 1977. In 1988 small canvases were distributed throughout the communities in the region, and some of the artists switched media, Kngwarray among them. Painting required less physical labor than melting wax and boiling fabrics, even as it offered a wider range of expressive possibilities while her eyesight was weakening. As the writer Jenny Green, herself a former community arts coordinator in Central Australia, has pointed out, that decade also saw a shift from an emphasis on communal work, such as that produced by batik, to works created by individual artists, which the contemporary art market prefers. The timing of her rise to prominence could not have been more fortuitous.

Kngwarray's dot paintings were at first laid over complex networks of lines, which related both to her landscape and the Anmatyerre yam vines, the Dreaming from which her middle name *Kame* is derived. She was mapping her ancestral territory, but also painting an aerial view of the vine with its yellow flowers and seeds—the dots—overlaying it. As in batik, where it is impossible to erase mistakes, which must then be incorporated into the ongoing layers of wax and dye, Kngwarray made mistakes and adopted them as freehand gestures helping her develop a unique style. Sitting on the ground in the middle of her largest canvases, she worked from the center out and switching hands as she went, the direction in which she painted and the ways in which to orient the painting equally of no consequence. The artist Virginia King, in her dissertation comparing the practices of Kngwarray and the Southwestern American minimalist Agnes Martin, who was working at the same time, notes how Kngwarray turned what was a defect in batik into a successful working practice in painting.

The other thing to note about Kngwarray's paintings, as well as those by Johnny Warrangkula Tjupurrula and others using dots, is a surface movement perceived by the eye, a "shimmer" that Howard Morphy says signifies spirit living in the land, an energy present thus in both land and paintings. Kathleen Petyarre's work, large canvases based on the tracks of the Mountain Devil Lizard Dreaming, consist of dots arranged in lines that cross or often meet near the center of the canvas. In some of her works the stippling is so fine as to become mesmerizing for the viewer, as it often is for the painters who have to come to realize that the art market prizes most highly those paintings showing evidence of the most careful detailing. That's in contrast to traditional Aboriginal practice, which treasures the fidelity of the painted Dreaming to that of their stories.

The most recent paintings I've been looking at are the remarkable canvases by Doreen Reid Nakamarra, an artist who lives in remote Western Australia, but who maintains close ties with Papunya. Born in 1955, Reid started painting in 1996 and has progressed from relatively simple geometric designs to intricate grounds built up out of fine dots arranged in lines in two or three tonal variations of the same color. Seen hanging on a wall, the optical effect of her work is like looking at irregular horizontal bands of herringbones across which the eye refuses to rest. The ambivalence of the viewer's gaze, never able to sort out exactly what's up or down, creates a shimmer from masses of short lines akin to that achieved by Kathleen Petyarre with her stippled arrays. But once Nakamarra's canvases are laid out on the ground, you begin to perceive them as ridges and gullies running parallel with one another, and that is precisely what she is painting, the sandhills where she lives. The paintings are minimalist and abstract, yet maximalist in density of gesture and reference, not a map but an intense re-creation of natural form. There is depth to them that is almost palpable, and I can't help but think once again that the haptic roots of the acrylic paintings in the sand paintings has resurfaced.

After sitting on a bench and taking notes from the Bardon film, and contemplating one last time both the early Papunya and Namatjira paintings, Kim gives me a ride back to the motel, and then in the early evening I walk over to Pam's house to see how Kim is doing on her map for the Tanami trip. She has it spread out in the studio, a rectangle taking up most of the floor, almost 8 x 12 feet of canvas that she's primed with ochre paint. She's traced out the western cartographic grid on it to match the topo printouts from government maps, and outlined the main features for an area totaling a million-and-a-half acres, filling in the lakes and streams with a deep sky-blue, salt lakes with a thick white, longitudinal dunes with orange streaks. It's not designed to be topographically exact, but to act as an armature upon which people can put place and family names, a device around which to tell stories of their lives, as well as any traditional ones they'd want to share, all of which will be recorded with their consent.

Kim gestures in the air before her as if manipulating a ball of space. "Once they look at this and figure out where they are, then they get excited and sit down and tell stories." The map, when finished, will be a long scroll of land covered with dots of different colors, size and density applied by the painters to indicate different type of country and vegetation, a hybrid object that reverses the usual ratio of objectivity to subjectivity apparent in a Western topographical map versus a painting. All maps are subjective, everything from the choice of color to represent geographical features to the point size assigned towns is a matter of choice. It may be logical to make a lake blue, but the density of that color on a topo map versus the green for forest is a subtle privilege forced upon your attention. You see the water first, which in turn tells you something about its status as an economic resource in the society that produced the map. Paintings, of course, are taken to be subjective as a matter, beginning with the vantage point of the artist and the framing of the subject, be that a landscape or a nude.

In Kim's map the supposedly more objective topographical signs could get buried under the dots, which is to assume that the contour lines will be subjected to painterly matters. Yet the dotting would perhaps tell a traveler as much or more about the nature of the country and the journey across it than would the contour lines, even though the placement of the dots is done with an aesthetic in mind as much as to indicate anything about the land.

This Tanami map represents Mongrel Downs, and is part of a personal project to record traditional Warlpiri knowledge relating to the country where she grew up. She has also been asked to make a map with the Walmajarri, custodians of the northern part of the Canning Stock Route. Unlike paintings, these are meant to be cross-cultural documents with topographical features recognizable to Euro-Australians, but including Aboriginal "place names, Dreaming tracks, birth and conception sites, traditional campsites and ancestral names." Kim writes in an essay about her trip that the process "records people's stories, oral histories and traditional knowledge, using video and simply-produced large-format books. It gives access to material already gathered by professionals, archived in reports in the bowels of some agency or other, written in language that tests even my literacy."

Historian and curator Phillip Jones published a book recently that explores the mutual influences traded between Aboriginal and Colonial cultures through their artifacts. The title says it all, *Ochre and Rust*, red ochre being the ground pigment that "is a medium or agent of transcendence, from sickness to health, death to renewal, ritual uncleanness to cleanness, the secular to the sacred, the present reality to the Dreaming." Ochre is ferric oxide, natural rust. What we call rust is more commonly the ubiquitous mark of the machine age, the industrial cousin of ochre. Rust is what happens when iron-based metals and entropy collide; they go in the opposite direction via decay, from health to death. Jones estimates that museums and collections around the world may hold 250,000 Aboriginal artifacts from the colonial period in Australia, objects representing perhaps 10,000 transactions between Europeans and the original owners of those objects. What happened in most cases was trade, and, as Jones notes, when you look at what we consider Aboriginal art objects made after initial contact, often you find a mixing of traditional and indigenous materials, such as ochre, stone and bark, with imported string, glass, and metal.

The artifacts that Jones analyzes are as diverse as a whip made from a ceremonial Aboriginal club with knotted lashes added by an English sailor, a hafted Aboriginal ax made from metal abandoned by members of a doomed expedition, and even the early Christian art made by Albert Namatjira as tourist items. In most cases, the cross-cultural artifacts are evidence of conflict between Aboriginal and European cultures. Kim's maps are gestures meant in part to foster reconciliation among them. Once each map painting is finished, it's exhibited along with related paintings, photographs and the recorded stories. Kim poses this as a model for a process that could literally take place all over Australia.

"I imagine a series of maps shimmering across the roads and suburbs, the parks and farmlands and urban sprawl, recording the traces of indigenous knowledge, inscribed with fragments of all those lost languages surviving in the names of places, whispering to the country in the language it can understand. I see it as something in which all Australians would participate, a shared enterprise in which each contributes what they do best." It's this mutual process of discovery she thinks can integrate the past and future, and one that employs an aerial view of the world constructed from two very different cultural traditions. I would argue, of course, that both the bird's-eye view Europeans have used to map new terrain, and the representation of Dreamings arise from the same cognitive processing abilities, albeit to different yet slowly converging aims.

"What they do best," Kim says, meaning the whitefella laying out the grid and topographically consistent contours of the country, the blackfella animating that picture with knowledge of and reverence for its resources, its stories, its mysteries. Everyone loves a map, I think, standing up from looking at Kim's painting to discover that Pirate has picked up my reading glasses and happily chewed them into oblivion.

There's a nice story that the historian Kitty Hauser, who has written about both aerial photography and the aerial imagination in Aboriginal painting, recounts about the anthropologist Charles Mountford watching women at Warupuju dance a Dreaming. They used their feet in the sand in such a way that they made marks that were a drawing of the sandhill country in which they were dancing. Mountford actually published a picture of the "gutter-like marks" juxtaposed with one taken by C. T. Madigan from four thousand feet over the Simpson Desert. Hauser wonders if our technological prosthesis—aerial photography—isn't just a way to regain some power of aerial visualization that we've lost over time.

Hauser notes that aerial photos make apparent patterns on the ground that are sometimes invisible without a vantage point from above them, which may help explain how Aboriginal people, when put down in terrain new to them, can still predict where to find water and other resources, or to navigate without so much as looking at the sun. She wonders if this is due to their aerial imagination, and I would support that. Anthropologist Mike Smith, in noting that the Aborigines have an aerial sensibility from looking at the ground over which they constantly walk, postulates that as hunter-gatherers they have much more acute powers of observation about terrain than we do. It's a required survival skill reinforced through constant practice. They would know what winds blow when and from what direction, how water has flowed for centuries over a variety of land-forms, what kinds of rock respond in what specific ways to geomorphological forces. The Inuit with whom I have worked in the Arctic have the same awareness while on ice. Aboriginal people have what we would consider almost mystical powers of navigation, but it's at least based in close observation coupled with an ability to mentally rotate space in their minds. They fly over Australia, airplanes or not.

Furthermore, place that ability to manipulate space mentally within a performance culture of more than forty thousand years, one that uses mutually reinforcing media to encode a sense of place. Deploy dance, storytelling, song, body decoration, ground painting, and rock art to ensure such knowledge not only is preserved, but expanded over time. And then, as a kind of safety, tie your kinship rules to all of it, so that bloodlines reinforce both preservation of geographical boundaries and negotiations over them. Make the ownership of your totemic stories, the ones for which you personally are responsible, a matter of life and death, not just in terms of survival knowledge, but the basis for your legal system. Make the penalty for abuse of stories, such as one person painting another's Dreaming without permission, subject to ritual punishment.

Christine Nicholls, when writing about the paintings of Kathleen Pet-yarre, could be describing the entire culture of the Aborigines. "These paintings offer an *integrated* [her emphasis] spatial, environmental, economic, spiritual and moral 'reading' of the land, of Anmatyerr spatial history, if you like." In addition to this condensation of both physical and abstract spatial properties into a painting, she adds that each painting is accompanied by an oral narrative that can take hours to relate.

The adoption of the aerial imagination by everyone in Australia from the Aborigines of the desert to painters such as Sidney Nolan and photographer Ruby Davies—it all has to do with living right, with a response to a difficult landscape. And part of making landscape mysterious is a healthy response to it. That is, art is effective in terms of engaging us with the land not just by revealing ground truth—not just by revealing patterns and giving us pretty pictures to admire—but also by concealing, by making mysterious, by making us invest attention and time in looking at it. And that's a survival skill as useful in our current times as in the Paleolithic.

# 17. Over Canberra . . . and Los Angeles

**Balloons rely on the temperature differential between hot and cold** air to obtain lift, which is why balloon races and rides are held at dawn, and why my visiting partner Karen Smith and I are sitting in a van with several other couples at 7:00 A.M. by Canberra's Parliament House. Neither of us has ever been in a balloon before, it's Karen's birthday, and almost my last day in Australia after three months—you get the idea. What better way to mark the multiple occasions than with a flight over a city designed from a theoretical aerial vantage point and thus begging to be seen from above.

One of the ground crew lets go of a small black balloon that rises quickly and straight into the early morning air. This is not good, as we want to float north and over Lake Burley Griffin, which is right behind us, and we need a breeze. The lake was constructed in 1963 at roughly the center of the capitol, which is where the architects Walter Burley Griffin and Marion Mahony Griffin had placed it in their design for the city. The lake is roughly seven miles long by three-quarters of a mile wide, and has been a joy for me to see each day while living in the city. But without a breeze there's no way to maneuver the balloon over it, and to make the usual flight across the water.

We drive over to West Park at the far end of the lake and loft another trial balloon. The air is still, the sun not yet up, and a light ground fog clings in small depressions here and there. This time the indicator drifts ever so slightly over the lake, enough that our pilot, Alan Shore, deems we can fly. The crew lays out the large wicker basket on its side, pulls out the balloon from the truck to stretch it out on the grass, then turns on an air blower to open up the envelope. They ask several of us to grab the rim of the opening as the bag begins to inflate, and we peer inside as a space the size of a large ballroom begins to open. With your head inside, it's like watching a three-dimensional stained glass window stand up from the ground and you can't help but smile.

Once the entire length of the fabric has some air in it Alan fires up the three burners and it begins to rise. The envelope, basket, and burners are all made by Kavanaugh, Australia's only manufacturer of balloons, and this is one of their larger models. The envelope holds three hundred thousand cubic feet, the colorful craft rising into the air as tall as a multistory office building. While I'm hanging onto the balloon, Karen is behind it taking pictures. At the roar of the burners igniting, a large kangaroo bursts out of the bushes next to Karen and bounds away, startling her. Canberra is a city filled with parks and surrounded by hills, and the roos come into the city during droughts to munch on the grass. How can you not love a city that has kangaroos and parrots?

At 7:35 we lift off—perhaps lurch off is more accurate? We rise a few inches, drift down, bounce, rise up, and gently we're then up and away. Alan radios air control to let them know we're launched. A smaller balloon from another company takes off to our north a few hundred yards. Alan gets us up a couple hundred feet and then we promptly lose buoyancy as we enter warmer air. Alan blasts us back up. The cold air follows the contours of the land, staying low like water, a contour that becomes apparent as you float through it. There's a faint southwest breeze and it pushes us out over the lake where we greet the sun. For the first time in three months I'm able to appreciate the fact that Canberra is in a bowl— and there's an inversion to prove it. The bits of ground fog are being joined by a light scrim of smog as morning traffic builds up, the top boundary of the pollution making that contour line visible. A flight of sulphur-crested cockatoos passes underneath us, their white wings catching the light.

It's not below freezing, but we're all bundled in hats and gloves. Because balloons fly with the movement of the air, there's no wind from our passage, but it's chilly and I'm glad for the warmth from the burners

on my bare hands as I take notes. Alan pulls on straps to rotate us, opening the "speed vents" used worldwide. Balloons haven't changed much
since the 1700s, and as Alan puts it: "There aren't many moving parts. The
fabric is better now, and lasts longer. The biggest improvement is the vent-
ing system that Phil Kavanaugh invented. It allows you to rotate the balloon and it reseals quickly, but its biggest advantage is that it lets you land
quickly, to control the process a bit." This is a distinct advantage in an
urban area where you have to maneuver in between buildings, streets,
trees, telephone wires, power lines, and the occasional pedestrians gawk-
ing at you.

As we rotate and take in a full view of the city it becomes apparent how
complex the shoreline of the lake is as it winds through the city, and I
begin to grasp more fully what the maps and drawings by its architects
had proposed. Walter Burley Griffin was born in a suburb outside
Chicago in 1876, and as a young student was given permission by his parents to landscape their front yard. When he was sixteen he was already
drawing up town plans, and when he earned his bachelor's degree in
architecture from the University of Illinois, he continued to study urban
planning, such as it then existed. He also managed to squeeze in courses
about forestry and horticulture before returning to Chicago to enter practice. Chicago in the 1890s was an extraordinary place and time in which
to become an architect. Louis Sullivan was erecting the first steel-frame
buildings and uttering the famous words "form follows function," invent-
ing both the skyscraper and modernist architecture in the same breath.
His young acolyte Frank Lloyd Wright was shaping what would become
the Prairie School, and in 1893 what was then the largest world's fair
opened in the city. The World's Colombian Exposition was in essence the
first modern planned city, albeit a miniature one, and organized around
that progressive vision of urban planning known as City Beautiful, which
emphasized symmetry and unity in civic design.

In 1901, the same year that Griffin went to work in Frank Lloyd
Wright's Oak Park studio, the example of the Chicago exposition spurred
the U.S. Congress to begin realizing the full dimensions of Pierre Charles
L'Enfant's 1791 plan for Washington, D.C. Frenchman L'Enfant had
designed a capitol city of radial avenues and open spaces that both linked

local topographical features and seats of power, while allowing a series of intersecting diagonals to expand outward over a rectilinear urban grid. It was a design that encouraged horizontal vistas, and invited the incorporation of ceremonial spaces and monuments, but his vision had never been completed, and was instead compromised and truncated during the nineteenth century. Daniel Burnham, another Chicago practitioner of the skyscraper and designer of the Flatiron Building in New York City, was considered the preeminent architect at the turn of the century, and had been director of works for the exposition. Burnham was brought to Washington and along with Frederick Olmsted, Jr., began to redesign the Mall that now frames the Capitol. They had the railroad tracks running down the middle of it ripped out and a park put in, and created unified spaces for monuments and government buildings to anchor the city. What they had in mind was what they had seen in their fact-finding visit to Paris early in the design process—a city of broad boulevards radiating out from a great public space into neighborhoods. In 1906 Burnham was hired to do the same thing in Chicago. He set aside the Lakefront area as off-limits to industrial development and created the parklands we see there today, proposed a civic center that is now the Loop, and had broad boulevards radiate out from it toward the very suburbs in which Griffin had been born and in which he was working.

The young architect Griffin had a falling out with Wright that year and went into practice for himself, but in 1911 married one of Wright's more important designers and the country's first licensed female architect, Marion Mahony. While on their honeymoon they heard of a competition offered by the Australian government to design a new capitol city, and they launched into one of the most legendary feats of architectural imagination in the history of the discipline. Neither of them had ever been to Australia, much less the proposed site for Canberra, but in three months they generated an exquisite series of topographical maps, contour plots, site plans, and illustrations for a design of a city that owed its inspiration if not directly to L'Enfant, then to his idealism and the current re-envisioning of cities as low profile, high density urbanscapes woven together by radial boulevards, parklands, and ceremonial spaces. But Walter's passion for and experience in landscape design, and Marion's ability to synthesize that design into what may have been the most beautiful set of architectural drawings seen in the twentieth century, allowed them to push their ideas far beyond the Beaux Arts–style of the City Beautiful and Prairie School architects.

They started with the geography of the site, creating a land axis between two hills and crossing it at right angles with a water axis that would consist of parks and a lake created by damming part of the Molonglo River. They then devised an equilateral triangle linking three areas of concentric streets, one each for federal, commercial, and municipal buildings. Marion's drawings of the plan, the buildings and boulevards, arrived in Australia as an immense set of folding panels eight feet wide and thirty feet long. Her imagined aerial perspective of the planned city as if painted from atop Mount Ainslie, the nearby peak from which the central axis of the city would be designed, is nothing less than astonishing in the breadth and accuracy of its panoramic siting in the landscape. The complete set of drawings was so striking that the judges had miniatures of them made for consideration, lest they simply award the prize to Griffin based on Marion's drawings, versus his design (as if the two were separate constructs). It's important to note that the earliest aerial photograph made in Australia, a bird's-eye view of the Sydney Harbor taken from a dirigible, wouldn't be made until 1919.

The Griffins received word in 1912 that they'd won the commission and Walter was able to visit Canberra for the first time the next year, whereupon he made a number of adjustments to the plan. Local politics, World War I, and any number of revisionist fingers in their design caused the Griffins to leave the project in 1920, but they had fallen in love with the country and remained here until Walter's death in 1937, whereupon Marion returned to the United States.

Alan, in rotating the balloon, has allowed us to observe how Parliament House sits atop the low brow of Capitol Hill to our right and faces Mount Ainslie, at the foot of which stands the Australian War Memorial. That's the land axis. The lake axis is what we are following in the balloon. The concentric circles of City Center, the commercial hub, are to our left now, as is the Australian National University and the apartment in which I've been living. As we float over the lake, I ponder how Aboriginal paintings, with their geometrical arrangements of circular sacred sites connected by lines, are related to the plan of Canberra. Perhaps resonate is a more accurate verb. The system-reticulum of Aboriginal Australia, the network created by the Dreaming tracks, codifies heritage and law by linking landmarks into paths that can cross the entire continent. The symbolic civic design and architecture of individual buildings in Canberra functions similarly at a symbolic level for the Euro-Australians. They both, the Dreaming tracks and boulevards, make physically manifest a system of belief and governance. And they are both represented by designs made from an aerial perspective.

The Aboriginal people, however, may not see that in Canberra as much as they perceive a landscape of loss. Anthropologist Rhys Jones described the Anbarra elder Frank Gurrmanamana's reaction to visiting the city:

Here was a land empty of religious affiliation; there were no wells, no names of the totemic ancestors, no immutable links between land, people and the rest of the natural and supernatural worlds. Here was just a tabula rasa, cauterized of meaning. Discussing the history of this place and being shown archaeological sites and nineteenth-century pictures of old Aborigines of the region, Gurrmanamana said that once long ago, Aborigines had lived there and that they would have known these attributes of the land which still existed somewhere, but that now, in his own words 'this country bin lose 'im Dreaming'.

It's a melancholy thought, the replacement of one system of memorialization by another, and that fact is acknowledged in three public artworks. From the balloon I can see in the forecourt of Parliament House a large mosaic by the prominent Papunya artist Michael Nelson Tjakamarra. *Possum and Wallaby Dreaming* is designed as a dot painting executed in ninety thousand granite chips. It covers 9,859 square feet in the typically muted desert palette. Across the lake is the National Museum of Australia, its courtyard another kind of outdoor ground map, the *Garden of Australian Dreams,* an artwork that lays out Australia as a stylized jigsaw of language groups. Lines crossing the concrete map represent both Euro-Australian and Indigenous boundaries. Another graphic element of the museum is a broad red line that begins at the front of the building, a vivid stripe aligned as if emanating from the Parliament House across the water, extending through the Aboriginal galleries, piercing the wall, and then running along the front of the Australian Institute of Aboriginal and Torres Strait Islander Studies. A sidewalk at that point, it proceeds in the direction of Uluru, Ayers Rock in European terms. It ends as a concrete ramp that points up and northwest toward Uluru, but also curls like a wave back upon itself, as if any attempt to link the two is destined to failure. The skateboarders like it either way.

The *Uluru Line*, as the three-dimensional super-graphic is known, makes a counterpoint to the land and water axes of Griffin's civic layout, being offset roughly 45° degrees from it and oriented to a landscape feature not in the valley, but rather one in the middle of the continent. It's an effort to tie together metaphorically the Dreaming tracks and the design of Canberra, to acknowledge that the two systems bear a relation to one another in human intent, as well as politics. Making the original Aboriginal occupation of Australia visible is not a gesture with which everyone is comfortable, however. Many of the exterior aluminum panels have words spelled out in silver disks set as braille; among them the phrases "sorry" (the code word necessary for reconciliation between whites and blacks in Australia) and "forgive us our genocide." As my sister Sue and I had found out when exploring the garden with a docent, the government of John Howard had demanded that the words be obscured, which the architects did by translating the into the alphabet of the blind, before the building opened. From our perch inside the wicker basket, the *Uluru Line* is the brightest color visible this morning. Being literally built into the building and landscape, the line was beyond the power of the government to obscure.

Alan has managed to get us drifting toward the northern shore of the lake, and it's a lovely and almost surreal moment when we sink below the height of the office buildings and, while clearing power lines and gum trees, can sight down a broad avenue just above the traffic. The ground crew has been following us and is waiting in a small park to see if Alan can maneuver his way over power lines and in between the trees to land.

We're so low over the passing cars that I'm amazed we don't cause any pileups, but we cross the street as if in a dream, at the speed and level with the trees and buildings I often find myself envisioning when airborne in my sleep. We just clear the tops of the foliage and settle on the grassy field. After we empty the envelope and get it stuffed back into its bag, we drive back over the hotel where we'd met, and debrief over the traditional glass of champagne. Legend has it that the custom derives from one of the first balloon flights in France, when the pilot had to calm a frightened and angry farmer with some bubbly lest the suspicious local puncture the balloon. Or the pilot.

Alan Shore was born in Kent in 1946 into a dairy farmer's family. He was a restless soul, and by the early 1970s was leading treks in Nepal. We briefly compare notes, as I'd been doing the same thing that decade, but while I kept my day job as an arts administrator in Nevada, he moved to South Africa where he took his first balloon ride and was immediately hooked. He joined a club, signed up for twelve hours of instruction, became a licensed balloonist, and eventually graduated to being an instructor himself. He migrated to Australia in 1984, and ten years ago started Dawn Drifters, his commercial ballooning company. Australia is a country amenable to the sport, given its relatively flat terrain, and balloons were here as early as the 1850s. Today about one hundred thousand people take rides annually in the lighter-than-air vehicles.

Alan also competes successfully in balloon races internationally, and he walks me through the changes. The winners used to be decided by how close they could land to a selected location. That evolved into dropping bags from the basket to hit a ground target. Now the competition is based on the ability of the pilots using a GPS unit to fly through a set of coordinates in the air. I mull over the purity of that concept while Karen and I drive back to the apartment and begin to pack for the long transcontinental flight back home. It's not about your relationship to the ground, but how you are able to locate and fly into a completely invisible point hovering in space. There's no sense of scale or touch, nothing to taste or hear or smell that can help you. It's just pure flight, another example of making an air place out of space.

## . . . AND LOS ANGELES, AGAIN

I wanted to end my travels in aereality by going up again with Mike Light to look at Los Angeles, this time at night, specifically to compare it with the experience of being over Canberra. The character of Canberra, given its distinctive and compact design, is more visible during the day than L.A., the grid of which reveals itself more easily at night through street-lights. And it will be a good chance to complete a circuit in my thoughts after having had a chance to work through some of the cognitive issues about aerial images. So a few weeks after my return from Australia, I'm climbing into a helicopter at the Van Nuys airport in the San Fernando Valley. Once again Mike has rented an R-44, and we load up a little after 7:00 on a July evening. We'll be flying with the doors off, but won't even need a jacket tonight, it's so warm.

I'm a little skittish because a few hours earlier two news helicopters had collided over Phoenix while covering a car chase. Both pilots and their cameramen had died. The pictures taken by passersby on the ground of the two aircraft smoking and plummeting toward the ground from five hundred feet were horrific. Five TV copters and one police chopper were in the air over the scene, an all-too-familiar situation today in large cities where car chases drive up the broadcast ratings. It was the first incident of its type in the U.S., and perhaps only the second worldwide. Witnesses say the aircraft simply got too close and, in essence, "sucked each other in." In L.A. it seems as if I'm seldom out of earshot of a helicopter, they are so ubiquitous by the major freeway interchange where we live, and under the flight path for aircraft passing from the San Fernando Valley to the L.A. Basin.

We're flying with ORBIC, a helicopter service specializing in aerial film shoots based at Van Nuys, which is the world's busiest general aviation airport. Our pilot is Tim Lerma, an ex-Marine medic of Hawaiian-Filipino-German descent. He's been flying helicopters for three-and-a-half years, and prior to that was jumping out of choppers for four years in places like Fallujah. He decided to learn how to fly when one chopper he was in had its engine quit and the pilot was able to land it safely. "Cowboys," I'm thinking to myself, " . . . they're all cowboys." My friend Larry Conrad now pilots commercial jetliners for a living, but before that he was a U.S.

Navy helicopter pilot who flew in, among other places, the Antarctic. Larry is as brave as anyone I know, but recently admitted to me: "A helicopter is a collection of parts flying in loose formation around an oil leak," apparently a common sentiment among pilots of the somewhat delicate aircraft. If the helo pilots have a certain amount of bravado about them, they're perhaps even more conscious than their colleagues flying fixed-wing aircraft of how precarious an endeavor human flight is outside our dreams.

Mike, of course, just ignores the issue of Phoenix and the fickle nature of the rotary aircraft, and he scarfs down a gigantic fast food hamburger before getting his camera ready. He's hustling to get us into the air right after the photographer's magic hour, just as the sun has gone down but with a trace of light left to define the horizon. We take off an hour later, just after 8:00, Tim lifting up and backwards in that counterintuitive motion so different from the forward thrust of a fixed-wing craft, yet so familiar now in a helicopter. The sun set twenty minutes ago, and there's an almost a full moon rising on this Friday night. It's a little hazy down low, but in essence we have unlimited visibility once we're up a few hundred feet. The terminator, the great vaulting shadow of the earth, rises in the east and curves up into the first stars.

Tim sets course over Interstate 405 and we pass over the Sepulveda Basin, a flood overflow catchment that's dry at this time of year, although there's a trickle of water in the L.A. River below the dam. He keeps over the freeway as we lift south over the Hollywood Hills toward the Getty complex of museums and institutes. From 1,500 feet we rise out of the relatively bounded basin of the San Fernando Valley and peer over into the much larger Los Angeles Basin. It's a very different perception than driving in between the two on Mulholland, much more three dimensional by comparison. The atmosphere is a soft blue-gray, liquid silver moonlight pouring over everything. Below us the northbound lines of the freeway are narrowed by an accident, the highway patrol directing traffic. It would be frustrating to be stuck down there on such a nice evening, and we have a feeling of freedom, cruising L.A. on a Friday night with no traffic. I'm beginning to enjoy myself.

Actually, correction. There's a lot of traffic up here, albeit nothing close to us. The airspace over L.A. appears huge tonight, as if we're swimming in an enormous aquarium. Helicopters and private planes are buzzing around Van Nuys and Santa Monica, the two general aviation facilities, plus there are the commercial jets coming from and going into both Burbank and LAX. I count more than twenty lights in the sky over the two basins. There's so much volume of open space around us, however, that it's peaceful.

We circle the Getty, its travertine gleaming in lights, a white castle. Despite my weekly work in the library there, I'm still surprised by how unified the architecture and gardens are from the air. Mike points Tim downward into the L.A. Basin toward another bright source of illumination, which turns out to be the lit tennis courts at UCLA. He's attracted like a moth to any source of light. I'm listening to Tim report into flight control, and realize that as he passes from one designated airspace to another and switches towers, he gives our position in relationship to buildings on the ground—first the Getty, then UCLA, now the Hugh Hefner Playboy Mansion. Just a little different from flying over Fallujah. Our flight traverses Beverly Hills and we follow Wilshire Boulevard into downtown, most of the windows in the skyscrapers lit up. As we pass over Frank Gehry's Disney Hall, it appears more as a landform than as a building, given its steep curvilinear forms that are so unlike the obsessive orthogonal grids of streets and steel-frame buildings around it. Mike has Tim fly us around downtown several times as he shoots the thousands of gleaming windows, and I have to hand him a new magazine of film.

As artists have known all along, the great horizontal grid of L.A. is made more visible by darkness. According to the Bureau of Street Lighting, the city has 5,000 miles of lit streets with more than 240,000 streetlights. The older and light-polluting mercury vapor models, which cast a greenish-white glow, are being replaced with low pressure sodium fixtures, their more modest orange light shielded to hit the ground and not the sky. As a result, the contrast between lines and bounded areas is more pronounced. The grid glows like lines of hot coals in a banked fire as far as I can see, the smaller street blocks encapsulated in larger ones defined by boulevards, those tracts in turn subsumed into the freeway network. The lines of power articulate a civic machine that is less about governance, as in Canberra, and more about the movement of people, information, and goods. It's more nakedly about money (Plate 10).

What a difference this fractal grid is from the curvilinear design of Canberra. Los Angeles is an accretion driven by developers tiling block to block, versus the planned design of Canberra with its concentric circles connected in a symbolic framework. You can decipher information about the history and socioeconomics of L.A. from the air, as we did with Denis Cosgrove over Long Beach and Lakewood, but the level of encoding practiced in Australia by both Aborigines countrywide and Euro-Australians in their capitol seems to be missing here. Ed Ruscha's photographs of parking lots and deadpan streetmap paintings, James Doolin's elevated soaring freeway overpasses, the painter Peter Alexander's nighttime views, and Florian Maier-Aichen's infrared panoramas begin to create an aerial identity of sorts for the city, one based on the value of the encompassing, synoptic view that humans desire of their landscape, be it through bird's-eye views, panoramas, or Google Earth. But Canberra is a capitol city, and the seats of government are places where we deliberately concentrate both power and metaphor. L.A. is a kind of anti-capital, its horizontality symbolizing, if anything, the diffusion of power versus the concentration of it.

While we're circling downtown and I'm scribbling, Tim is chatting away, far more at ease in the air than the other helicopter pilots with whom we've flown. He's precise when he follows Mike's requests, instinctively tilts the chopper so the blades are out of the frame of the camera, and never loses track of other aircraft or conversation with air traffic control—but he's also having fun. For one thing, he grew up in Glendale just north of downtown, and knows the city intimately from the ground. Plus, he flies television news crews around every afternoon from 2:00 to 6:00 P.M., so he knows terrain from the air well. I ask him about Phoenix and he shakes his head. "It's hard to fly and report," he replies.

Mike asks him if it's possible to fly in between the office towers, and Tim replies in the negative. Only senior pilots are allowed permission to make the riskier flights. "I'm working the bottom of the ladder here, flying as a location scout for people. The last one I did was for Acura, looking around for highway curves with a natural smile for an ad campaign.

I really want to do movie work, which pays the best, but you have to put in your time to get there. I've been doing bank courier flights, too. It's a lot cheaper to fly millions around than to use a truck."

As I'm digesting this piece of monetary trivia, I'm thinking about the ubiquity of Hollywood worldwide and envisioning armored helicopters, how the nighttime view of downtown L.A. is now familiar to hundreds of millions, if not billions, of people around the world. And how global banking could be envisioned as a connected and fractal grid like the one below us—and then it strikes me. Anti-capital, indeed. Power, both political and physical, and money are both increasingly decentralized, dispersed, and distributed through a series of interconnected grids on paper, in person, in electrons; and L.A. is the perfect city to embody that devolution. The writer and columnist Thomas Friedman has been talking about how the world is going "flat," and this view manifests what he means. If the design of Canberra seeks to symbolize a traditional and centralized European model of authority, L.A. exemplifies the growth of a grid that knows no boundaries and self-replicates to the horizon in the global economy.

We head for the next big light, Dodger Stadium, where much to our surprise the field is littered with sleeping bags and people camped-out watching baseball videos on a giant screen. Tim brings us around low in a circle so Mike can photo the spectacle from different angles, probably much to the annoyance of the spectators. Only in L.A.: a camp-out in the middle of baseball stadium being photographed from a helicopter. Tim peels away from the rim and as we head toward Mulholland, takes care to avoid the Hollywood Bowl. "See the strobes there, two of them on the hills above the Bowl?" It's a 'noise sensitive zone,' and the lights let us know there's a concert and to stay away." That's how many helicopters there are flying at any given time over the city—a concert venue has to post signs warning them away.

As we follow the dark and mostly invisible Mulholland, the helicopter flies through small pockets of cooler air. "It's like diving in the ocean," Mike comments and smiles. I nod, realizing how much more comfortable I've become in copters, knowing that part of my accommodation stems from getting used to—even addicted to—that feeling of quickly and fluidly changing your body's position in a three-dimensional medium. Your sense of how your body moves in space begins to loosen and you feel empowered to assume a place in aereality. It's presumptuous, I remind myself, whether it's from the mechanical standpoint, or the carbon footprint created. Icarus, I think. He would make a perfect, if decidedly unpopular, saint for helicopter pilots.

We're back on ground about 9:20. Mike's taken 450 or so photos, and we're both a bit dazzled from looking down at so many lights. As always, I'm wondering what images Mike will make from the experience. His last L.A. night photos devolved down into isolated pinpricks of light on a black background, as if the dark city were the night sky. I think about that reversal as we drive home, turning what should be familiar ground into a mysterious constellation. It's as if, given that we are born with an aerial imagination, Mike and other aerial photographers who make the world anew from the air are telling us not to take the ability for granted, or to think that just because we can see over the world that we are capable of actually overseeing it.

It is exactly the opposite of what I thought of aerial views when I started thinking about them. I proposed at first that the aerial view of Çatalhöyük was the work of an individual genius, and that Leonardo's elevated views of northern Italy confirmed that supposition. Wrong. It's a widespread human ability extending across the world's cultures. And I thought the purpose of flying high to see the world was a mapping one, when in fact it is as much about metaphor and mystery. But, then, that's the purpose of writing a book, at least for a writer, just as it is for a geographer to climb the next ridge: to find out if what you think is true.

*When once you have tasted flight, you will forever walk the earth with your eyes turned skyward, for you have been there and there you will always long to return.*

—**Leonardo da Vinci,**
**as quoted by the painter Hung Liu**

# Acknowledgments

Books written over a number of years and set in so many locations invariably require lengthy acknowledgments because the writer has relied upon a great deal of help from others. Such is the case here.

What I define as my home library is the Getty Research Institute, and many people there have supported this work, most prominently Tom Crow and Charles Salas. Charles and I organized a symposium at the "GRI" in spring 2007 titled "The God's Eye View," which brought together Denis Cosgrove, Michael Light, Lucia Nuti, and myself to present aerial images and discuss them. It was, I admit with some embarrassment, a gathering of people who had shaped most profoundly my inquiry into matters aerial. Lucia is an historian of urban planning and an expert on the development of bird's-eye city views in Europe, and thus embodies the connection of art to cartography. Denis, who died just as I was completing this book, was a revered historian of cartography teaching at UCLA whose scholarship included that same conjunction of images. Mike is, of course, the artist with whom both Denis and I had been flying as he photographed, and his images are one of the shared passions that brought Denis and I together in the first place. The symposium was the first time I presented publicly some of the ideas in this book, and I wish to thank all of the above for their patient support and guidance. The loss of Denis as a friend and colleague is immeasurable. His work in that very special place where cartography, geography, and art come together is foundational for anyone hoping to navigate such terrain.

Thanks are also due to other GRI staff people: Carol Casey, Cathy Davis, Jay Gam, Roberta Panzanelli, and Sabine Schlosser. One memorable evening the year prior to the symposium, Roberta introduced me to the great Leonardo da Vinci scholar Carlo Pedretti over dinner, which proved to be an important event for this work. Weston Naef, the curator of photographs at the Getty Museum, has been a convivial informant to my investigations, as have his staff members Anne Lyden, Judy Keller, and Virginia Heckert.

During the summer of 2006 the Clark Art Institute in western Massachusetts hosted me as a visiting fellow, and I happily wrote the first chapters in its library. Michael Ann Holly, Gail Parker, and Karen Bechy were most excellent hosts. It was also at the Clark that I met the curator Thierry Davila, who provided the title for this book. That summer Joe Thompson from the Massachusetts Museum of Contemporary Art was kind enough to take me flying and show me some of the differences between western and eastern American aereality.

My time in Australia for researching the "Down Under" portion of the book was made possible by a visiting fellowship from the Humanities Research Centre (HRC) at the Australian National University. My profound thanks to Ian Donaldson, former director of the HRC, for suggesting that I apply, and to his wife, the internationally renowned curator and scholar Grazia Gunn, who introduced me to much of Australia's modern and contemporary art. Leena Messina at the HRC has taken care of all the fellows for years with patience and humor, and we all owe her a bow.

No sooner was I off the plane in Canberra than the environmental historian Bernadette Hince and her husband, novelist Nick Drayson, took me in hand. Mandy Martin and Guy Fitzhardinge, who are friends to many artists and writers worldwide, had me out to Pennyroyal, their property in the central tablelands of New South Wales, while Kim Mahood drove me over two thousand miles of Outback roads. John Carty, even as he was initiating me into the intricacies of the Aboriginal art market, lent me his swag for the trip. All of them are now friends with whom I continue to work on other projects.

Among my many other colleagues in Canberra and elsewhere helping me toward a preliminary understanding of Australia were Bill Gammage, Mike Gillam, Tom Griffiths, Pam Lofts, Helen Maxwell, Kim Mackenzie, John Reid, Libby Robin, Deborah Bird Rose, Mike Smith, and Carolyn Strange. Wally Caruana, Howard Morphy, and Cath Bowdler were kind enough to help me approach Aboriginal art, and should be held harmless for any mistakes I make regarding that field of study. Martin Woods and Damien Cole at the National Library invited me into that institution's astonishing collection of aerial photographs and maps, and Ruby Davies took me around Sydney to see various instances of aerial art.

Upon my return from Australia, the anthropology and earth sciences departments at California State University, Dominguez Hills, hosted me as their first distinguished visiting scholar on campus, throughout which are planted groves of eucalypt. The fragrance of their oils kept me walking on the wrong side of the sidewalk, much to the amusement of my American colleagues. The CSUDH anthropologist and archeologist Jerry Moore, a former fellow scholar at the Getty, was responsible for the idea, and he remains a role model for me of field researcher, writer, and teacher.

While in residence at CSUDH, Denis Cosgrove and I continued to work on a collaboration for Reaktion Books in London, *Photography and Flight*, an idea that Denis had suggested to the publisher the previous year as a book that he and I might write. That corollary to *Aereality* was one of the most satisfying books upon which I've had the pleasure to work either solo or with others. I sketched out the historical and cognitive aspects of the subject, then Denis reshaped the material to sharpen its arguments, and shaded in the contours of social history. That work informs every level of *Aereality*; I am humbled to have worked with a scholar of Denis's abilities and warmth.

The generosity, skill, and time given me by the pilots with whom I have flown over the years can't be praised enough, and I thank them in the various chapters. But there is one person who was key, and who wasn't present as I began my research. I had several memorable flights in a Maule tail-dragger over much of the desert West with Blanton Owen during the 1980s, when he was the Folk Arts Coordinator at the Nevada Arts Council. Blanton went down over Washington State several years later, but not before setting me on the course to write this book, although neither of us knew it at the time.

Several years ago, while working on my book *Playa Works*, Matt Coolidge of the Center for Land Use Interpretation (CLUI) hired a pilot to fly us out of Overton, Nevada, up to Michael Heizer's "Double Negative," and then to look for Walter de Maria's "Las Vegas Piece." Matt and CLUI have also been kind enough to put both Michael Light and myself up at the organization's residence in Wendover, Utah, which sits on the edge of the tarmac and across from the World War II *Enola Gay* hangar, a primary piece of American aerial architecture. I've been fortunate to work in Wendover over the years with several other writers, artists, and architects, all of whom have broadened my understanding of land and landscape, among them Bill Gilbert and Chris Taylor of the Land Arts of the American West program, artists Katherine Bash and Lucy Raven, Steve Rowell and Eric Knutzen from CLUI, and photographer Mark Klett.

In every writer's life there is a group of friends who continually lend their enthusiasm and beachcombing abilities to the work. David Abel, Robert Beckmann, and Bruce McAllister are always sending arcane bits of flotsam that they run across, which sometimes turn out to be pivotal bits of information.

The publisher Jack Shoemaker, senior editor Roxanna Aliaga, and Trish Hoard for her fine editing skills.

Victoria Shoemaker is my agent, friend, patient counselor, and all-purpose sanity provider. Bless you, Victoria.

And then there is Karen Smith, my partner, travel companion, first reader, and close friend during the writing of this book. Her one goal throughout it was, who knows why, to get a ride in a helicopter.

# Selected Bibliography

## Aerial Imagery/Photography

Adams, Ansel, and Nancy Newhall. *This Is the American Earth*. San Francisco: Sierra Club, 1960.

Aldrich Museum of Contemporary Art. *Landscapes of Consequence*. Ridgefield, CT, 1991. Exhibition brochure with ground-based photographs by Robert Glenn Ketchum and Richard Misrach, and aerials by David T. Hanson and David Maisel.

Alexander, Peter. *Peter Alexander: In This Light*. Essays by Dave Hickey and Naomi Vine. Newport Beach, CA: Orange County Museum of Art, 1999.

Apt, Jay, Michael Helfert, and Justin Wilkinson. *Orbit: NASA Astronauts Photograph the Earth*. Washington, DC: National Geographic Society, 1996.

Arthus-Bertrand, Yann. *The Earth from the Air*. Paris: La Martinière, 1999.

Baker, Simon. "San Francisco in Ruins: The 1906 Photographs of George R. Lawrence." *Landscape*, Vol. 30 (1989).

Baudrillard, Jean. *America*. Translated by Chris Turner. London: Verso, 1988.

Berger, Alan. *Reclaiming the American West*. New York: Princeton Architectural Press, 2002.

———. *Drosscape: Wasting Land in Urban America*. New York: Princeton Architectural Press, 2006.

Blankenship, Edward G. *The Airport*. New York: Praeger Publishers, 1974.

Blaut, James M. "Natural Mapping." *Transactions of the Institute of British Geographers*, New Series, Vol. 16, no. 1 (1991): 55–74.

Bourke-White, Margaret. *Portrait of Myself*. New York: Simon & Schuster, 1963.

———. *Twenty Parachutes*. Tucson: Nazraeli Press, 2002.

Breydenbach, Bernhard von. *Peregrinatio in Terram Sanctam*. Incunabula: In ciuitate Moguntina: Opusculum hoc co[n]tentiu[m] p[er] Erhardu[m] reüwich de Traiecro inferiori impressum, Anno salutis 1486, die xj Februarij. Mainz: Erhard Reuwich, 1486.

Bridges, Marilyn. *Markings: Sacred Landscapes from the Air*. New York: Aperture, 1986.

————. *The Sacred and Secular: A Decade of Aerial Photography*. Essay by Vicki Goldberg. Interview by Anne H. Hoy. New York: International Center for Photography, 1990.

Brigidi, Stephen, and Claire V. C. Peeps. *Mario Giacomelli*. Carmel, CA: Friends of Photography, 1983.

Brown, Julia, ed. *Michael Heizer: Sculpture in Reverse*. Los Angeles: Museum of Contemporary Art, 1984.

Brown, Theodore M. *Margaret Bourke-White: Photojournalist*. Ithaca, NY: Andrew Dickson White Museum of Art, Cornell University, 1972.

Bunnell, Peter C. *Emmet Gowin: Photographs 1966–1983*. Washington, DC: Corcoran Gallery of Art, 1983.

Burkhard, Balthasar. *Photographer*. Zurich: Scalo, 2004.

Callahan, Sean, ed. *The Photographs of Margaret Bourke-White*. New York: Bonanza Books, 1972.

————. *Margaret Bourke-White: Photographer*. Boston: Little, Brown, 1998.

Campanella, Thomas J. *Cities from the Sky: An Aerial Portrait of America*. New York: Princeton Architectural Press, 2001. More than 100 photographs taken from the 1920s to 1960s by the Fairchild Aerial Survey Company are accompanied by texts from Witold Rybcyinski and————.

Casey, Edward S. *Representing Place: Landscape Paintings and Maps*. Minneapolis: University of Minnesota Press, 2002.

Castleberry, May. *The New World's Old World: Photographic Views of Ancient America*. Albuquerque: University of New Mexico Press, 2003.

Celant, Germano. *Michael Heizer*. Milan: Fondazione Prada, 1997.

Clark, Kenneth, and Carlo Pedretti. *The Drawings of Leonardo da Vinci in the Collection of Her Majesty the Queen at Windsor Castle*. Vols. 1–2. London: Phaidon, 1969.

Connell, David, ed. *Uncharted Waters*. Canberra: Murray-Darling Basin Commission, 2002.

Cooke, Lynne, and Karen Kelly, eds. *Robert Smithson: Spiral Jetty*. Berkeley: University of California Press, 2005.

Corner, James, and Alex S. MacLean. *Taking Measure Across the American Landscape*. New Haven, CT: Yale University Press, 1996. This collaboration between noted aerial photographer MacLean and landscape architect Corner, who contributed essays, commentary, maps, and drawings, also includes an essay by Denis Cosgrove. It signaled a shift to the use of aerial images that integrated landscape theory and archeology with land use and urban planning, based not just on technical appraisals but also on ground aesthetics as perceived from the air.

Cosgrove, Denis. *Apollo's Eye*. Baltimore: Johns Hopkins University Press, 2001.

Cosgrove, Denis, and William L. Fox. *Photography and Flight*. London: Reaktion Books, 2008.

Denby, Edwin, ed. *Aerial*. New York: Eyelight Press, 1981. Small anthology of aerial poems with aerial images by Yvonne Jacquette.

Dicum, Gregory. *Window Seat: Reading the Landscape from the Air*. San Francisco: Chronicle Books, 2004. An armchair guidebook to deciphering aerial views.

Diesel, Eugen, and Robert Petschow. *Das Land der Deutschen*. Leipzig: Bibliographisches Institut AG, 1931.

Donnelly, Erin, Moukhtar Kouche, Olu Oguibe, Liz Thompson, Anthony Vidler et al. *Site Matters: The Lower Manhattan Cultural Council's World Trade Center Artists Residency, 1997–2001*. New York: Lower Manhattan Culture Council, 2004.

Dreikausen, Margaret. *Aerial Perception: The Earth as Seen from Aircraft and Spacecraft and Its Influence on Contemporary Art*. Philadelphia: The Art Alliance Press, 1985.

Drutt, Matthew. *Olafur Eliasson: Photographs*. Houston: Menil Foundation in association with D.A.P., 2004.

Dunaway, Finis. *Natural Visions: The Power of Images in American Environmental Reform*. Chicago: University of Chicago Press, 2005.

Eldredge, Charles C. *Georgia O'Keeffe: American and Modern*. New Haven: Yale University Press, 1993.

Evans, Terry. *Prairie: Images of Ground and Sky*. Lawrence: University Press of Kansas, 1996.

———. *Disarming the Prairie*. Baltimore: Johns Hopkins University Press, 1998.

———. *The Inhabited Prairie*. Lawrence: University Press of Kansas, 1998.

———. *Revealing Chicago: An Aerial Portrait*. New York: Harry N. Abrams, 2006.

Faberman, Hilarie. *Aerial Muse: The Art of Yvonne Jacquette*. New York: Hudson Hills Press, 2002.

Frank, Robert. *The Americans*. New York: Scalo, 1993 (originally published in America by Grove, 1959).

Fried, William. *New York in Aerial Views*. New York: Dover Publications, 1980.

Garcia Espuche, Albert. *Cities: From the Balloon to the Satellite*. Madrid: Centre de Cultura Contemporania de Barcelona/Electa, 1994.

Garnett, William. *The Extraordinary Landscape: Aerial Photographs of America*. Introduction by Ansel Adams. Boston: Little, Brown, 1982.

———. *William Garnett: Aerial Photographs*. Introduction by Martha A. Sandweiss. Berkeley: University of California Press, 1994. While the former title by Garnett is a color book of predominantly nature photographs from the air, this second book is a highly aestheticized presentation of black-and-white art images based on strong ground patterns perceptible primarily, or even exclusively, from the air.

Gerster, Georg. *Flights of Discovery*. New York: Paddington Press, 1978.

———. *Brot und Salz: Flugbilder*. Basel: Birkaeuser Verlag, 1985.

———. *The Past from Above*. Oxford: Oxford University Press, 2005.

Gohlke, Frank. *Landscapes from the Middle of the World*. San Francisco: Friends of Photography, 1988.

Gosling, Nigel. *Nadar*. London: Secker & Warburg, 1976.

Gowin, Emmet. *Emmet Gowin: Photographs*. Essay by Martha Chahroudi. Philadelphia: Philadelphia Museum of Art, 1990.

————. *Changing the Earth*. New Haven: Yale University Press in conjunction with the Corcoran Gallery of Art. This exhibition catalog includes essays by exhibition curator Jock Reynolds and nature writer Terry Tempest Williams ("The Earth Stares Back"), and an interview with Philip Brookman.

Greenough, Sarah, Robert Gurbo, and Sarah Kennel. *André Kertész*. Washington, DC: National Gallery of Art, 2005.

Grysztejn, Madeleine, Daniel Birnbaum, and Michael Speaks. *Olafur Eliasson*. New York: Phaidon, 2002.

Hall, Charles. *Paul Nash: Aerial Creatures*. London: Imperial War Museum, 1996.

Halsey, Adriel. *Under the Sun: A Sonoran Desert Odyssey*. Tucson: Rio Nuevo, 2000.

Hanson, David T. *Waste Land: Meditations on a Ravaged Landscape*. New York: Aperture, 1997.

Harley, J. B., and David Woodward, eds. *The History of Cartography*. Chicago: University of Chicago Press. The first volume of this magisterial and definitive series was published in 1987 and contains a seminal essay on early mapping images, including the wall mural at Çatalhöyük, by Catherine Delano Smith, "Cartography in the Prehistoric Period in the Old World: Europe, the Middle East, and North Africa." From that same volume I also consulted P. D. A. Harvey's discussion of early bird's-eye city images in "Local and Regional Cartography in Medieval Europe."

Hauser, Kitty. *Shadow Sites: Photography, Archaeology, and the British Landscape 1927–1951*. Oxford: Oxford University Press, 2007.

Hawkes, Jason. *Aerial: The Art of Photography from the Sky*. East Sussex, UK: Rotovision, 2004. Images and text by a leading UK commercial photographer.

Hayden, Dolores. *A Field Guide to Sprawl*. New York: W. W. Norton, 2004.

Hickey, Dave, and Naomi Vine. *Peter Alexander: In This Light*. Newport Beach, CA: Orange County Museum of Art, 1999.

Hickson, Patricia. "James Doolin's Illusionistic Space." In *Urban Invasion: Chester Arnold and James Doolin*. San Jose: San Jose Museum of Art, 2001.

Hodder, Ian, ed. *On the Surface: Çatalhöyük 1993–95*. Cambridge: McDonald Institute for Archaeological Research, 1996.

————. *The Leopard's Tale: Revealing the Mysteries of Çatalhöyük*. London: Thames & Hudson, 2006.

Howat, John K. *Frederic Church*. New Haven, CT: Yale University Press, 2005.

Ingleby, Richard, et al. *C. R. W. Nevinson: The Twentieth Century*. London: Merrell Holberton, 2000.

Johnson, George R., Lt. *Peru from the Air*. New York: American Geographical Society, 1930.

Kennedy, Marla Hamburg, and Ben Stiller, eds. *Looking at Los Angeles*. New York: Metropolis Books, 2003.

Kertész, Andre. *Kertész on Kertész*. New York: Abbeville Press, 1985.

Kimmelman, Michael. "A Sculptor's Colossus of the Desert." *New York Times*, December 12, 1999.

———. "Art's Last, Lonely Cowboy." *New York Times*, February 6, 2005.

Kost, Julieanne. *Window Seat: The Art of Digital Photography and Creative Thinking*. Sebastopol, CA: O'Reilly, 2006.

Lartigue, Jacques-Henri. *Boyhood Photos of J. H. Lartigue: The Family Album of a Gilded Age*. Lausanne: Amy Guichard, 1966.

Le Corbusier. *Aircraft: The New Vision*. London: The Studio Ltd., 1935.

Light, Michael. *Ranch*. Santa Fe: Twin Palms, 1993.

———. *Full Moon*. New York: Alfred A. Knopf, 1999.

———. *100 Suns: 1945–1962*. New York: Alfred A. Knopf, 2003.

Light, Richard Upjohn. *Focus on Africa*. Photographs by Mary Light. New York: American Geographical Society, 1944.

Lopez, Barry. "Flight." In *About This Life*. New York: Alfred A. Knopf, 1998.

Lynes, Barbara Buhler. *Georgia O'Keeffe: Catalogue Raisonné*. New Haven: Yale University Press, 1999.

MacDonald, Angus, and Patricia MacDonald. *Above Edinburgh and South-East Scotland*. Edinburgh: Mainstream Publishing, 1989. A physical and cultural geography by a husband-and-wife team (she photographs, he pilots and writes) that is an informed look at how geology and history have shaped the city.

MacLean, Alex, with text by Bill McKibben. *Look at the Land: Aerial Reflections on America*. New York: Rizzoli, 1993.

———. *Designs on the Land: Exploring America from the Air*. London: Thames & Hudson, 2003. A topology of historical and contemporary land usage patterns with an introduction by Corner and essays by MacLean, Gilles A. Tiberghien, and Jean-Marc Besse.

Malevich, Kasimir. *The Non-Objective World*. Trans. Howard Dearstyne. Chicago: Paul Teobold and Company, 1959.

Maisel, David. *Exposed Terrain*. New York: David Maisel Photography, 1990. Early work by Maisel after he photographed the aftermath of the Mt. Saint Helens explosive eruption.

———. *The Lake Project*. Tucson: Nazraeli Press, 2004.

———. *Oblivion*. Portland, OR: Nazraeli Press, 2006.

*The Manual of Photogrammetry*, 2nd ed. Washington, DC: American Society of Photogrammetry, 1952.

Marinoni, Augusto, ed. *Leonardo da Vinci: The Codex on the Flight of Birds in the Royal Library at Turin*. Foreword by Carlo Pedretti. New York: Johnson Reprint Corporation, 1982.

Marshall, Richard D. *Ed Ruscha: Made in Los Angeles*. Madrid: Museo Nacional Centro de Arte Reina Sofia, 2002.

———. *Ed Ruscha*. London: Phaidon Press, 2003.

Martin, Rupert, ed. *The View from Above: 125 Years of Aerial Photography*. London: The Photographers' Gallery, 1983.

McGill, Douglas C. *Michael Heizer: Effigy Tumuli*. New York: Harry N. Abrams, 1990.

Mellaart, James. *Çatalhüyük: A Neolithic Town in Anatolia*. London: Thames and Hudson, 1967. This is the popular account of the original excavation by Mellaart (hence the different spelling of the site name). Although many of Mellaart's interpretations have since been disproved, in particular his insistence on the prevalence of goddess figures, it contains valuable insights and illustrations.

Monmonier, Mark. "Aerial Photography at the Agricultural Adjustment Administration: Acreage Controls, Conservation Benefits, and Overhead Surveillance in the 1930s." *Photogrammetric Engineering and Remote Sensing*, Vol. 68, no. 11 (December 2002): 1257–1261.

Morris, Robert. *Continuous Project Altered Daily*. Cambridge: MIT Press, 1995.

Morse, Rebecca. *Florian Maier-Aichen*. Los Angeles: Museum of Contemporary Art, 2007.

Naef, Weston. *In Focus—Andre Kertész: Photographs from the J. Paul Getty Museum*. Malibu, CA: J. Paul Getty Museum, 1994.

Nash, Paul. *Aerial Flowers*. Oxford: Counterpoint Publications, 1947.

Newhall, Beaumont. *Airborne Camera: The World from the Air and Outer Space*. New York: Hastings House, 1969.

Nuti, Lucia. "The Perspective Plan in the Sixteenth Century: The Invention of a Representational Language." *The Art Bulletin*, Vol. 76, no. 1 (March 1994): 105–128.

Nye, David E. "The Sublime and the Skyline." In *The American Skyscraper*, ed. Roberta Moudry. Cambridge: Cambridge University Press, 2005.

Oesterreichische Galerie Belvedere. *The Waste Land: Desert and Ice. Barren Landscapes in Photography*. Vienna: 2001.

Paschel, Huston, and Linda Johnson Dougherty. *Defying Gravity: Contemporary Art and Flight*. Raleigh: North Carolina Museum of Art, 2003. An exhibition catalog with contributions by Robert Wohl and others.

Pascoe, David. *Airspaces*. London: Reaktion Books, 2001.

Pauli, Lori. *Manufactured Landscapes: The Photographs of Edward Burtynsky*. Ottawa: National Gallery of Canada, 2003.

Peters, Sarah Whitaker. *Becoming O'Keeffe: The Early Years*. New York: Abbeville Press, 1991.

Phillips, Sandra S., David Travis, and Weston J. Naef. *Andre Kertész of Paris and New York*. Chicago: Art Institute of Chicago, 1985.

Price, Alfred. *Targeting the Reich: Allied Photographic Reconnaissance over Europe, 1939–1945*. London: Greenhill Books, 2003.

Pyne, Stephen J. "SPACE: The Third Great Age of Discovery." In *SPACE: Discovery and Exploration*. New York: Hugh Lauter Levin, 1993.

Reps, John. *Bird's-eye Views: Historic Lithographs of North American Cities*. New York: Princeton Architectural Press, 1998.

Reynolds. Ann. *Robert Smithson: Learning from New Jersey and Elsewhere*. Cambridge: MIT Press, 2003.

Ristelhuber, Sophie. *FAIT: Kuwait 1991*. Paris: Editions Hazan, 1992.

———. *Details of the World*. Boston: Museum of Fine Arts, 2001.

Roth, Matthew. "Mulholland Highway and the Engineering Culture of Los Angeles in the 1920s." In *Metropolis in the Making Los Angeles in the 1920s*, ed. Tom Sitton and William Deverell. Berkeley: University of California Press, 2001.

Ruscha, Edward. *Leave Any Information at the Signal: Writings, Interviews, Bits, Pages*. Cambridge: MIT Press, 2000.

———. *Ed Ruscha: Photographs*. Beverly Hills, CA: Gagosian Gallery.

Schaaf, Larry J. *The Photographic Art of William Henry Fox Talbot*. Princeton: Princeton University Press, 2000.

Schulthess, Emil. *Antarctica*. New York: Simon & Schuster, 1960. An early book by the pioneer of modern European aerial photography, who photographed extensively both on the ground and in the air over Russia, Africa, and the Alps, and who also served as the art director for Swissair.

———. *Swiss Panorama*. Text by Emil Egli, translated into English by A. J. Lloyd. New York: Alfred A. Knopf, 1983.

Sekula, Allan. "The Instrumental Image: Steichen at War." *ArtForum*, December 1975, pp. 26–35.

Sichel, Kim. *To Fly: Contemporary Aerial Photography*. Boston: Boston University Art Gallery, 2007.

Siegel, Beatrice *An Eye on the World*. New York: Frederick Warne & Co., 1980.

Smithson, Robert. "Towards the Development of an Air Terminal Site." *Artforum* 6 (June 1967): 36–40.

———. "Aerial Art." *Studio International* 177 (April 1969): 180–181.

St. Joseph, J. K. S., ed. *The Uses of Air Photography: Nature and Man in a New Perspective*. New York: John Day Company, 1966.

Städtische Galerie im Lenbachhaus München. *Olafur Eliasson: Sonne statt Regen*. Ostfeldern: Hatje Cantz, 2003.

Stanley, Roy M. Col., *World War II Photo Intelligence*. New York: Scribner's Sons, 1981.

Sweetman, John, Brandon Taylor, and Jonathan Bayer. *The Panoramic Image*. Southampton: John Hansard Gallery, 1981. Exhibition catalog.

Turrell, James. *Air Mass*. London: South Bank Centre, 1993.

———. *Infinite Light*. Scottsdale, AZ: Scottsdale Museum of Contemporary Art, 2001.

Varanka, Dalia. "Interpreting Map Art with a Perspective Learned from J. M. Blaut." *Cartographic Perspectives*, no. 53 (Winter 2006): 15–23.

Waldheim, Charles. "Aerial Representation and the Recovery of Landscape." In *Recovering Landscape: Essays in Contemporary Landscape Architecture*, ed. James Corner. New York: Princeton Architectural Press, 1999.

Walsh, Michael J. *C. R. W. Nevinson: This Cult of Violence*. New Haven: Yale University Press, 2002.

Weems, Jason. "Aerial Views and Farm Security Administration Photography."
*History of Photography*, Vol. 28, no. 3, (Autumn 2004): 267–282.
Whimster, Rowan. *The Emerging Past: Air Photography and the Buried Landscape.*
London: Royal Commission on the Historical Monuments of England, 1989.
Whitfield, Peter. *Cities of the World: A History in Maps.* Berkeley: University of
California Press, 2005.
Wohl, Robert. *A Passion for Wings: Aviation and the Western Imagination.* New
Haven: Yale University Press, 1994.
———. *The Spectacle of Flight: Aviation and the Western Imagination, 1920–1950.*
New Haven: Yale University Press, 2005.
Wolf, Sylvia. *Ed Ruscha and Photography.* New York: Whitney Museum of Ameri-
can Art, 2004.
Yochelson, Bonnie. *Berenice Abbott: Changing New York.* New York: Museum of
the City of New York, 1997.
Zukowsky, John. *Building for Air Travel: Architecture and Design for Commercial
Aviation.* Chicago: Art Institute of Chicago, 1996.

## Australia

Bardon, Geoffrey. *Aboriginal Art of the Western Desert.* Adelaide: Rigby Ltd., 1979.
———. *Papunya Tula: Art of the Western Desert.* Ringwood, Victoria: McPhee
Gribble, 1991.
Bardon, Geoffrey, and James Bardon. *Papunya: A Place Made After the Story.* Carl-
ton: The Miegunyah Press, 2004.
Bonyhady, Tim. *Burke and Wills: From Melbourne to Myth.* Sydney: David Ell
Press, 1991.
Bonyhady, Tim, and Tom Griffiths, eds. *Landscape & Language in Australia.* Sydney:
University of New South Wales Press, 2002.
Bruce, Candice, Edward Comstock, and Frank McDonald. *Eugene von Gerard,
1811–1901: A German Romantic in the Antipodes.* Martinborough, New Zealand:
Alister Taylor, 1982.
Caruana, Wally. *Aboriginal Art.* London: Thames & Hudson, 2003. Along with
Howard Morphy's book, the standard and excellent overview of the
subject.
Commonwealth of Australia. *Report from the Senate Select Committee Appointed to
Inquire and Report upon the Development of Canberra.* Appendix B, 93–102,
"Copy Federal Capital Design No. 29 by W. B. Griffin. Original Report." Sep-
tember, 1955.
Connell, David, ed. *Uncharted Waters.* Canberra, Murray-Darling Basin Commis-
sion: 2002.
Coyne, Peter. *Protecting the Natural Treasures of the Australian Alps.* Canberra: Aus-
tralian Department of the Environment and Water Resources, 2001. A report
to the Natural Heritage Working Group of the Australian Alps Liaison

Committee. A copy online can be found at http://www.australianalps.deh. gov.au/publications/natural-treasures/pubs/natural-treasures.pdf.

Darragh, Thomas A. "Ludwig Becker, a Scientific Dilettante: His Correspondence with J. J. Kaup and Others." *Historical Records of Australian Science*, Vol. 11, no.4 (December 1997): 501–522.

Davidson, Robyn. *Tracks*. New York: Vintage, 1980.

————. *No Fixed Address: Nomads and the Fate of the Planet*. Quarterly Essay, no. 24 (2006).

Davies, Ruby. *The Darling*. Sydney: Self-published chapbook, 2001.

————. *Contested Visions, Expansive Views: The Landscape of the Darling River in Western NSW*. Sydney: University of Sydney, Sydney College of the Arts, 2006. Accessed 1/30/08 at http://ses.library.usyd.edu.au/handle/2123/1119.

Grant, Kirsty, and Jennifer Phipps. *Fred Williams: The Pilbara Series*. Melbourne: National Gallery of Victoria, 2002.

Griffiths, Tom. *Forest of Ash: An Environmental History*. Cambridge: Cambridge University Press, 2001.

Hart, Deborah, *John Olsen*. Roseville East, New South Wales: Craftsman House, 1991.

Hauser, Kitty. "On Seeing and Old Country from the Air." Lecture given at the Humanities Research Centre, Australian National University, Canberra, September 2005.

Haynes, Roslynn D. *Seeking the Centre: The Australian Desert in Literature, Art and Film*. Cambridge: Cambridge University Press, 1998.

Hudjashov, Georgie, et al. "Revealing the Prehistoric Settlement of Australia by Y Chromosome and mtDNA Analysis." *Proceedings of the National Academy of Sciences of the United States of America (PNAS)*, Vol. 104, no. 21, (May 22, 2007): 8726–8730.

Jacobs, Michael. *The Painted Voyage: Art, Travel and Explorations, 1564–1875*. London: British Museum Press, 1995.

Johnson, David. *The Geology of Australia*. Cambridge: Cambridge University Press, 2004.

Jones, Philip. *Ochre and Rust: Artefacts and Encounters on Australian Frontiers*. Kent Town, South Australia: Wakefield Press, 2007.

King, Virginia. "Art of Place and Displacement: Embodied Perception and the Haptic Ground." Ph.D. dissertation, unpublished, 2005. Accessed January 2008 at http://www.library.unsw.edu.au/~thesis/adt-NUN/uploads/approved/adt-NUN20060116.112542/public/02whole.pdf.

Lynch, Tom. "Literature in the Arid Zone." In *The Littoral Zone: Australian Contexts and Their Writers*, ed., C. A. Cranston and Robert Zeller. Amsterdam: Rodopi, 2007.

Mackay, Donald, and Herbert Basedow. "The Mackay Exploring Expedition, Central Australia, 1926." *The Geographical Journal*, Vol. 73, no. 3 (March 1929): 258–265.

————. "The Mackay Aerial Survey Expedition, Central Australia, May–June 1903." *The Geographical Journal*, Vol. 84, no. 6 (Ced., 1934): 511–514.

Madigan, C. T. "The Australian Sand Ridges." *Geographical Review*, Vol. 26, no. 2 (April 1936): 205–227.

Mahood, Kim. *Craft for a Dry Lake*. New York: Random House, 2001.

———. "Listening Is Harder Than You Think." *Griffith Review 19: Re-imagining Australia*. Brisbane: Griffith University, 2008.

Mahood, Marie. *Legends of the Outback*. Rockhampton, Queensland: Central Queensland University Press, 2005.

Martin, Mandy. *Tracts: Back O'Bourke*. Manurama, Australia: Mandy Martin and Paul Sinclair, 1996.

———. *Watersheds: The Paroo to the Warrego*. Manurama, Australia: Mandy Martin and Tom Griffiths, 1999.

———. *Inflows: The Channel Country*. Contributions from Jane Carruthers, Guy Fitzhardinge, Tom Griffith, and Peter Haynes. Manurama, Australia: Mandy Martin, 2001.

———. *Percipecia: The Salvator Rosa Series*. Contributions by David Malouf and Nancy Sever. Canberra: Drill Hall Gallery, Australian National University, 2002.

———. *Land$cape: Gold & Water*. Manurama, Australia: Mandy Martin. Collaborative research project in residence at Cadia Hill Gold Mine.

———. *Strata: Deserts Past, Present and Future*. Contributions by Libby Robin and Mike Smith. Manurama, Australia: Mandy Martin, 2005.

———. "Absence and Presence." In *Fresh Water: New Perspectives on Water in Australia*, ed. Emily Potter, et al. Melbourne: Melbourne University Press, 2007.

McCaughey, Patrick. *Fred Williams*. Sydney: Bay Books, 1980.

McGregor, Ken. *William Creek & Beyond*. Sydney: Craftsman House, 2002.

Millar, Ann. *I See No End to Travelling: Journals of Australian Explorers, 1813–76*. Sydney: Bay Books, 1986.

Mollison, James. *A Singular Vision: The Art of Fred Williams*. Canberra: Australian National Gallery, 1989.

Morphy, Howard. *Aboriginal Art*. London: Phaidon Press, 1998.

Mulvaney, John, and Johan Kamminga. *Prehistory of Australia*. Crows Nest, New South Wales: Unwin & Allen, 1999. This is the fundamental starting place for the topic.

Myers, Fred R. *Painting Culture: The Making of an Aboriginal High Art*. Durham: Duke University Press, 2002.

Nicholls, Christine. *Kathleen Petyarre: Genius of Place*. Kent Town, South Australia: Wakefield Press, 2000.

O'Brien, Gregory. "The Dark Plane Leaves at Evening." New Zealand Electronic Text Center, *Sport 21* (Spring 1998.) Accessed May 4, 2007, at http://www.nztec.org/tm/scholarly/tei-Ba21Spo-tl-body-d6.html.

Olsen, John. *The Land Beyond Time*. Perth: Art Gallery of Western Australia, 1984.

Pearson, Michael. *Great Southern Land: The Maritime Exploration of Terra Australis*. Canberra: Department of Environment and Heritage, 2005.

Phipps, Jennifer, and Kirsty Grant. *Fred Williams: The Pilbara Series*. Melbourne: National Gallery of Victoria, 2002.

Pyne, Stephen. *Burning Bush: A Fire History of Australia*. New York: Henry Holt, 1991.

Reed, Dimity, ed. *Tangled Destinies: National Museum of Australia*. Mulgrave, Victoria: Images Publishing Group, 2002.

Robin, Libby. *How a Continent Created a Nation*. Sydney: University of New South Wales Press, 2007.

Rose, Deborah Bird. *Nourishing Terrains: Australian Aboriginal Views of Landscape and Wilderness*. Canberra: Australian Heritage Commission, 1996. The Rhys Jones quote is found here.

Scott, John, and Richard Woldendorp. *Landscapes of Western Australia*. Bridgetown, WA: Aeolian Press, 1986.

Seddon, George. *Landprints: Reflections on Place and Landscape*. Cambridge: Cambridge University Press, 1997. In particular, "Journeys Through a Landscape."

*Site Seeing*. Alice Springs: IAD Press (Institute for Aboriginal Development), 1991.

Smith, Bernard. *European Vision and the South Pacific*. New Haven: Yale University Press, 1985.

Smith, Geoffrey. *Sidney Nolan: Desert & Drought*. Melbourne: National Gallery of Victoria, 2003.

Smith, Mike, and Paul Hesse, eds. *23°S: Archeology and Environment History of the Southern Deserts*. Canberra: National Museum of Australia Press, 2005. A superb hemisphere-wide context for Australia's natural and cultural prehistories.

Sutton, Peter. "Icons of Country: Topographic Representations in Classical Aboriginal Traditions" and "Aboriginal Maps and Plans." Both in *The History of Cartography*, Volume 2, Book 3: *Cartography in the Traditional African, American, Arctic, Australian, and Pacific Societies*. Chicago: University of Chicago Press, 1998.

Witter, Dan. "Aboriginal Dreaming and Aridity." In *Animals of Arid Australia*, ed. by Chris Dickman, David Lunney, and Shelley Burgin. Mosman: Royal Zoological Society of New South Wales, 2007.

Will Owens's remarkable website and blog, "Aboriginal Art & Culture: An American Eye" (http://homepage.mac.com/will_owen/iblog/index.html) presents "Readings, reviews, and reflections by an American observer of Australian Indigenous art, culture, politics, anthropology, music, and literature." My Australian colleagues remain amazed that he can cover his subject so well from so far away, a high compliment indeed.

## Cognition & Other

Ambroziak, Brian M., and Jeffrey R. Ambroziak. *Infinite Perspectives: Two Thousand Years of Three-Dimensional Mapmaking*. New York: Princeton Architectural Press, 1999.

Appleton, Jay. *The Experience of Landscape*. Chichester: John Wiley & Sons, 1996.

Armstrong, Carol. *Scenes in a Library*. Cambridge: MIT Press, 1998.

Aveni, Anthony F. *Between the Lines: The Mystery of the Giant Ground Drawings of Ancient Nasca, Peru*. Austin: University of Texas Press, 2000.

Basso, Keith H. *Wisdom Sits in Places*. Albuquerque: University of New Mexico Press, 1996.

Bateson, Gregory. *Steps to an Ecology of Mind: Collected Essays in Anthropology, Psychiatry, Evolution, and Epistemology*. Chicago: University of Chicago Press, 1972.

———. *Mind and Nature*. New York: E. P. Dutton, 1979.

Bednarik, Robert G. "The Earliest Evidence of Paleoart." *Rock Art Research*, Vol. 20, no. 2: 89–135. Bednarik is a controversial independent scholar who makes wide-ranging but serious claims about the antiquity of Australian rock art versus that of Europe.

Boettger, Suzaan. *Earthworks: Art and the Landscape of the Sixties*. Berkeley: University of California Press, 2002.

Cantile, Andrea, ed. *Leonardo genio e cartografo*. Florence: Istituto Geografico Militare, 2003. Exhibition catalog that accompanied show curated by Cantile, the subtitle of which in English is *The Representation of Land in Science and Art*.

Casey, Edward S. *Representing Place*. Minneapolis: University of Minnesota Press, 2002.

Cassati, Roberto, *The Shadow Club*. New York: Alfred A. Knopf, 2003.

Crary, Jonathan. *Techniques of the Observer*. Cambridge: MIT Press, 1991.

Davis, Mike. *City of Quartz: Excavating the Future in Los Angeles*. London: Verso, 1990.

———. *Ecology of Fear: Los Angeles and the Imagination of Disaster*. New York: Metropolitan Books, 1998.

Dissanayake, Ellen. *Homo Aestheticus: Where Art Comes from and Why*. Seattle: University of Washington, 1995.

Hodgson, Derek. "Art, Perception and Information Processing: An Evolutionary Perspective." *Rock Art Research*, Vol. 17, no. 1, (2000): 3–34. As are many people undertaking studies into the nature of petroglyphs and pictographs, Hodgson is an independent scholar speculating in an informed fashion about how recent research in cognition might illuminate the origins of artifacts labeled rock art.

Humboldt, Alexander von. *Personal Narrative of Travels to the Equinoctial Regions of the New Continent During the Years 1799–1824*. Vols. 1–6. Translated by Helen Maria Williams. London: Longman, Rees, Orme, Brown, and Green, Paternoster Row, 1826.

Kaplan, Ellen. *Achademia Leonardi Vinci*. Vol V, *Leonardo's Telltale Legacy, Buckminster Fuller to Jasper Johns*. 1992, 143–146.

Kastner, Jeffrey, ed. *Land and Environmental Art*. London: Phaidon, 1998.

Kerr, Laurie. "The Mosque to Commerce." Posted at Slate Online, December 28, 2001. http://www.slate.com/?id=2060207.

Kitchin, Rob, and Mark Blades. *The Cognition of Geographic Space*. London: I. B. Tauris, 2002.

Komanoff, Charles. "Whither Wind." *Orion*, September/October 2006, pp. 29–37.

Launius, Roger D. "Home on the Range: The U.S. Air Force Range in Utah, a Unique Military Resource." *Utah Historical Quarterly*, Vol. 59 (1991): 332–360.

Lindqvist, Sven. *A History of Bombing*. New York: The New Press, 2000.

Livingston, Margaret. *The Biology of Seeing*. New York: Harry N. Abrams, 2002.

MacEachren, Alan M. *How Maps Work: Representation, Visualization, and Design*. New York: The Guilford Press, 1995. A useful discussion of the cognitive and semiotic issues of how we visually represent geography.

Manguel, Albert, and Gianni Guadalupi. *The Dictionary of Imaginary Places*. Maps by James Cook. New York: Harcourt Brace 2000. The point that the only way to present them is through aerial perspective.

Novak, Barbara. *Nature and Culture: American Landscape and Painting 1825–1875*. New York: Oxford University Press, 1980.

O'Sullivan, John. "The Great Nation of Futurity." *The United States Democratic Review*, Vol. 6, 23 no. 426–430.

Panzanelli Clignett, Roberta. *Achademia Leonardi Vinci*. Vol. X, "Plasticum ante alia penicillo praeponebat.'" 1997, pp. 152–160.

Pedretti, Carlo. *Achademia Leonardi Vinci*. Vol. IX, "Mural Perspective as Cinemascope: Story-board to Production." 1996, pp. 87–98.

Pyne, Stephen J. "SPACE: The Third Great Age of Discovery." In *SPACE: Discovery and Exploration*. New York: Hugh Lauter Levin, 1993.

Reisner, Marc. *Cadillac Desert: The American West and Its Disappearing Water*. Rev. ed. New York: Penguin, 1993.

Roque, Oswaldo Rodriguez. "'The Oxbow' by Thomas Cole: Iconography of an American Landscape Painting." *Metropolitan Museum Journal*, Vol. 17 (1982): 63–73.

Sachs, Aaron. *The Humboldt Current*. New York: Viking Press, 2006.

Solso, Robert. *Cognition and the Visual Arts*. Cambridge: MIT Press, 1994.

Urton, Gary. *At the Crossroads of the Earth and the Sky: An Andean Cosmology*. Austin: University of Texas Press, 1982.

Virilio, Paul. *War and Cinema: The Logistics of Perception*. Translated by Patrick Camiller. London: Verso, 1989.

Waldie, D. J. *Holy Land: A Suburban Memoir*. New York: W. W. Norton, 1996.

Weiner, Jonathan. *Time, Love, Memory*. New York: Alfred A. Knopf, 1999.

Woodward, David. *Art & Cartography: Six Historical Essays*. Chicago: University of Chicago Press, 1987.

The Center for Land Use Interpretation offers invaluable resources for anyone studying anthropogenic effects on landscape in North America: www.clui.org.

One source of figures quoted for 2001 Bingham pollution was the Environmental Protection Agency's Enforcement & Compliance History Online (ECHO) database, http://www.epa.gov/echo/.

The other source was the testimony of Lexi Shultz, Legislative Director of the Mineral Policy Center before the House Resource Committee, Subcommittee on Energy and Mineral Resources, "The Toxic Release Inventory and Its Impact on Federal Minerals and Energy," September 25, 2003. Her testimony was originally available online at http://resourcescommittee.house.gov/archives/108/testimony/lexishultz.pdf, but appears to have been scrubbed. The same figures were nonetheless reported by Patty Henetz in the May 12, 2005, issue of the *Salt Lake Tribune* in an article titled "Utah Third Nationally in Release of Toxins."

*The 9/11 Commission Report*. Washington, DC: National Commission on Terrorist Attacks Upon the United States, 2004. The complete report and ancillary materials are archived at http://www.9-11commission.gov/.

Air travel statistics for the "West to East" chapter were drawn from the Airports Council International *World Report Traffic Report 2006*, accessed online 04/08/08 at http://www.aci.aero/aci/aci/file/World%20Report%202004/WorldReportJULY2007.pdf.

# Credits

## Photos

All images by Michael Light, David Maisel, and Terry Evans are reproduced courtesy of the artists.

Mandy Martin's painting *The Tailings Dam* is courtesy of the Roslyn Oxley9 Gallery in Sydney, Australia.

Fred Williams, *Winjana Gorge, Kimberley, I*: Metropolitan Museum of Art.

Eugene von Guerard, *North-east view from the northern top of Mount Kosciusko*: National Library of Australia.

## Epigraphs

Bachelard, Gaston. *Air and Dreams: An Essay on the Imagination of Movement*. Dallas, TX: Dallas Institute of Humanities and Culture, 1988. Translated by Edith R. Farrell and C. Frederick Farrell from the 1943 publication.

Thomas J. Campanella. *Cities From the Sky: An Aerial Portrait of America*. New York: Princeton Architectural Press, 2001.

Cole, Thomas. "Essay on American Scenery." *American Monthly Magazine 1*, (January 1836) pp. 1–12.

Davis, Jack. *Black Life*. St. Lucia: Queensland University Press, 1992.

Greenaway, Peter. *Le bruit des nuages* (*Flying out of this world*). Paris: Réunion des musées nationaux, 1992.

Jackson, John Brinckerhoff. *Discovering the Vernacular Landscape*. New Haven: Yale University Press, 1984.

Langeweische, William. *Inside the Sky: A Meditation on Flight*. NY: Vintage Books, 1998

Lerner, Ben. *Angle of Yaw*. Port Townsend, WA: Copper Canyon Press, 2006.

Markham, Beryl. *West with the Night*. San Francisco: North Point Press, 1983.

Murray, Les. "Equanimity" in *The People's Otherworld*. Sydney: Angus & Robertson, 1983.

Stein, Gertrude. *Picasso*. London: B. T. Batsford, 1938.